DNA
Changing Science and Society

The year 2003 marks the fiftieth anniversary of the discovery of the double helical structure of DNA. Since the original revelation by James Watson and Francis Crick, knowledge of the structure and function of DNA has dramatically changed science and society. This volume explores the dramatic impact that this discovery has had on our lives. Beginning with the story of the discovery of the double helix, the collection looks at DNA fingerprinting and its impact on forensic and legal medicine; the extraction of ancient DNA from archaeological and palaeontological remains; the complex role of DNA in the cause, detection and treatment of cancer; the debates surrounding the potential commercialisation of genetically modified crops; the emotive field of reproductive medicine; how the genetic basis of developmental language disorders is teaching us more about how humans communicate; and finally the ethical implications arising from the genetic knowledge encoded in our DNA.

THE DARWIN COLLEGE LECTURES

DNA

Changing Science and Society

Edited by *Torsten Krude*

CAMBRIDGE
UNIVERSITY PRESS

PUBLISHED BY THE PRESS SYNDICATE OF THE UNIVERSITY OF CAMBRIDGE
The Pitt Building, Trumpington Street, Cambridge, United Kingdom

CAMBRIDGE UNIVERSITY PRESS
The Edinburgh Building, Cambridge CB2 2RU, UK
40 West 20th Street, New York, NY 10011–4211, USA
477 Williamstown Road, Port Melbourne, VIC 3207, Australia
Ruiz de Alarcón 13, 28014 Madrid, Spain
Dock House, The Waterfront, Cape Town 8001, South Africa

http://www.cambridge.org

First published 2004

Printed in the United Kingdom at the University Press, Cambridge

Typeface Iridium 10/14 pt.   *System* LaTeX 2_ε [TB]

A catalogue record for this book is available from the British Library

ISBN 0 521 82378 1 hardback

Contents

Introduction

TORSTEN KRUDE

Department of Zoology, University of Cambridge

Fifty years ago, on 25 April 1953, a short paper by James Watson and Francis Crick was published in the journal *Nature*, entitled 'Molecular structure of nucleic acids: a structure for deoxyribose nucleic acid'. Its authors described a model of DNA with two anti-parallel helical strands and complementary base pairing. Their model was built on X-ray diffraction data, which were described in two accompanying articles in the same issue of *Nature*, one by Maurice Wilkins and colleagues and one by Rosalind Franklin and Raymond Gosling. Watson and Crick discussed the biological implications of their model a month later in a second paper, entitled 'Genetical implications of the structure of deoxyribonucleic acid', again in *Nature*. However, they had already concluded their initial publication with the far-reaching statement that 'it has not escaped our notice that the specific pairing we have postulated immediately suggests a possible copying mechanism for the genetic material'.

It became clear in the years that followed that this discovery had sparked a revolution in our understanding of biological processes. Major implications arose, not only for science, but for human society more broadly. Today, many aspects of our world have fundamentally changed due to our understanding of DNA's structure and function.

The year 2003 has seen many events celebrating the golden jubilee of this discovery, here in Cambridge, where Crick and Watson worked in the Cavendish Laboratory (and announced in a local pub that they had discovered the secret of life), and elsewhere. Darwin College Cambridge marked this event by making it the topic of its annual interdisciplinary series of lectures 'The Darwin College Lecture Series'. This, the eighteenth year, saw eight distinguished speakers from a range of disciplines explore the impact of our understanding of DNA on contemporary science and society. These eight lectures form the basis of the following essays.

The opening essay by **Aaron Klug** tells the story of the very discovery of the DNA double helix. Drawing on original laboratory notebooks and first-hand contact with the scientists involved, he details the research behind the breakthrough (from the early investigations by Maurice Wilkins at King's College London through the sorting out of the two DNA forms by Rosalind Franklin to the final model-building by James Watson and Francis Crick). Klug's incisive historical analysis is then extended to include a consideration of the initial, often hesitant, reception of the proposed structure by the scientific community, its subsequent confirmation by biochemistry and X-ray crystallography and the final demonstration of the validity of the biological principle of 'semi-conservative' replication.

Alec Jeffreys conveys a personal account of the origin and extensive development of genetic fingerprinting (a technique which he invented). This technology is based on detecting variable and highly individual pieces of DNA, providing a new dimension to biological identification. Already it has had a huge impact on resolving identification issues in our society in, for example, criminal investigations, paternity disputes and immigration challenges. He discusses how the underlying biological principles of these variable pieces of DNA can be used to study fundamental aspects of human genetic variability and heritable mutation.

Svante Pääbo discusses his pioneering experiments of extracting and studying ancient DNA from remains of dead or extinct organisms, excavated by archaeologists and palaeontologists or collected by zoologists. He focuses on the damaged condition of ancient DNA and the technical challenges that this damage raises for the contemporary analysis of it. He also, however, shows that such analysis has already succeeded in deciphering the genealogical relationships of extinct organisms, including moas, the marsupial wolf and, last but not least, the Neanderthals.

Ron Laskey discusses the central role of DNA in cancer, highlighting its relevance to both the cause and the treatment of this disease. He argues that cancer is essentially a disease of damaged DNA that results in the loss of regulated growth in the affected cells. In turn, the most important treatments for cancer are based on inflicting yet more damage on DNA, to such an extent that the damaged cells commit suicide. We are in a position now where our increasing knowledge of DNA can in fact be exploited to improve diagnosis and treatment of a wide range of cancers.

Malcolm Grant explores mechanisms that shape social attitudes towards advances in biotechnology, a range of disciplines that make use of the manipulation and recombination of DNA to create new analytical tools, or novel varieties of plants or animals. He focuses on the dynamics of the polarised contemporary public debate over the potential commercialisation of genetically modified crops in the UK. He argues, based on personal experience, that credible policy-making in science and technology today involves more than merely building bridges between divergent intellectual elites or educating the public in science. He emphasises the need to develop more open political processes that will enable people to contribute in intelligent ways to society's thinking about difficult issues.

Robert Winston addresses selected topics in reproductive medicine, which are significant in the light of our recent understanding of DNA. He highlights relevant scientific discoveries from embryological research involving DNA analysis, and discusses their implications not only for clinical applications in humans, but for society in general. He begins by discussing aspects of the human embryo that are relevant for understanding infertility in humans, including gamete development, the process of ageing in women and the failure of the embryo to implant into the uterus. A major focus of his essay is a discussion of in-vitro-fertilisation procedures, with a particular emphasis on pre-implantation diagnosis. Finally he discusses cloning and transgenic animals, and embryonic stem cells.

Dorothy Bishop discusses the contribution of genetic information encoded in our DNA to the formation of the unique human ability of language. She departs from the old polarised debate between linguists who postulate an innate language module and psychologists who deny the need for anything special to explain language specialisation, towards a more moderated and differentiated approach to the problem. Her essay critically describes genetic studies of children with specific language impairments. She argues that these studies provide a unique source of information for developing our understanding of the complex pathways from genes, through neurobiology, to behaviour.

In the concluding essay, **Onora O'Neill** discusses ethical issues arising from the use and control of knowledge about the information encoded in our DNA. She identifies fundamentally new ethically relevant questions concerning access to our genetic information and its personal, familiar or collective ownership. Finally, she addresses the problematic issue of how informed an individual has to be in order to give consent to, or refuse, highly complex uses of personal genetic information with its wider implication for contemporary medicine.

Each of these eight essays is self-contained, highly individual and a reflection of the author's style and personality. Together, they explore the ramifications of the discovery of the DNA double helix made fifty years ago and take these into very different and relevant spheres of modern life, both in science and in society.

I wish to thank Richard Henderson, who co-organised the eighteenth Darwin Lecture Series with me; William Brown, the Master of Darwin College; the Fellows and students of the College; and Joyce Graham – for their help, expertise and encouragement.

1 The discovery of the DNA double helix

AARON KLUG

MRC Laboratory of Molecular Biology, Cambridge

Fifty years ago, on 25 April 1953, there appeared three papers in the journal *Nature*, which changed our world picture. The structure of the DNA double helix, with its complementary base pairing, was one of the greatest discoveries in biology in the twentieth century. It was also most dramatic, since, quite unexpectedly, the structure itself pointed to the way in which a DNA molecule might replicate itself, and hence revealed the 'secret of life'. The structure was solved in the Cavendish Laboratory, Cambridge, by Francis Crick and James Watson, using X-ray diffraction data from fibres of DNA obtained by Rosalind Franklin at King's College, London.

This essay aims to tell the story of how this came to happen: the origin of the research on DNA, the early investigations by Maurice Wilkins at King's College, the sorting out of the two forms of DNA by Franklin, the wrong paths taken, the intervention of old rivalries from an earlier generation, and the final model-building by Watson and Crick to give the three-dimensional structure.

The initial, often hesitant, reception of the proposed structure, and its confirmation by biochemistry and by X-ray crystallography at King's College, will be described. Yet this remained a discovery in chemistry, until the biological principle of 'semi-conservative' replication was proved by Meselson and Stahl.

Finally, a very brief summary will be given of the results of fifty years of research on the complex 'molecular machines' which carry out the two main functions of DNA, replication of the molecule, and transcription (reading) of

I thank Richard Henderson for inviting me to give the Darwin Lecture forming the basis of this essay, and for encouraging me to use the opportunity to give a more detailed account than such occasions usually warrant. I thank Francis Crick, Maurice Wilkins and Raymond Gosling for helpful discussions.

The publisher would like to thank Mrs Jenifer Glynn and the Churchill Archives Centre for permission to publish extracts from Rosalind Franklin's laboratory notebooks.

the sequence of the DNA to produce the RNA coding for the protein product of a gene.

The transforming principle

In 1945 the Royal Society of London awarded its highest honour, the Copley Medal, to Oswald Avery of the Rockefeller Institute of New York for 'his contributions to knowledge of the chemical basis of the specific properties of bacteria, particularly to the types of pneumococcus'.

Fred Griffith had earlier discovered that a non-pathogenic mutant of pneumococcus could be transformed into a pathogenic form by the addition of an extract from cells of the pathogenic form which had been killed by heat. This finding led Avery and his colleagues, MacLeod and McCarty, to try to establish the chemical nature of the 'transforming principle' (nowadays we would say – 'factor') in the extract and in 1944 they showed it was a 'nucleic acid of the deoxyribose type' – that is, DNA. In his 1945 Anniversary Address to the Society, the President Sir Henry Dale said 'Here surely is a change to which, if we were dealing with higher organisms, we should accord the status of a genetic variation, and the substance inducing it – the gene in solution, one is tempted to call it – appears to be a nucleic type of the desoxyribose type. Whatever it be, it is something which should be capable of complete description in terms of structural chemistry.'

The President's careful wording reflects the fact that it was believed by many scientists that bacteria did not have genes as did plants and animals. Eight years later the President's challenge was answered: there was a complete description of the 3-D structure of DNA – what a chemist would call its configuration – the double helix, by Watson and Crick, using X-ray diffraction data from Franklin and Wilkins. This essay aims to describe how this came about. Much of the story has been told in parts, in Watson's book, in articles by him and Crick, in Wilkins' Nobel Lecture, in the books by Olby and Judson, but Franklin's scientific work has never been fully described, and I have therefore drawn on her notebooks, now in the Archives of Churchill College, Cambridge, to document it.

The solution of the 3-D structure (Plate I) depended on knowing the correct chemical structure of DNA, that is, how the links in the phosphate – deoxyribose sugar backbone were made and how the heterocyclic nitrogenous bases were connected to the sugar. This had been worked out only two years earlier by Brown and Todd (Figure 1.1).

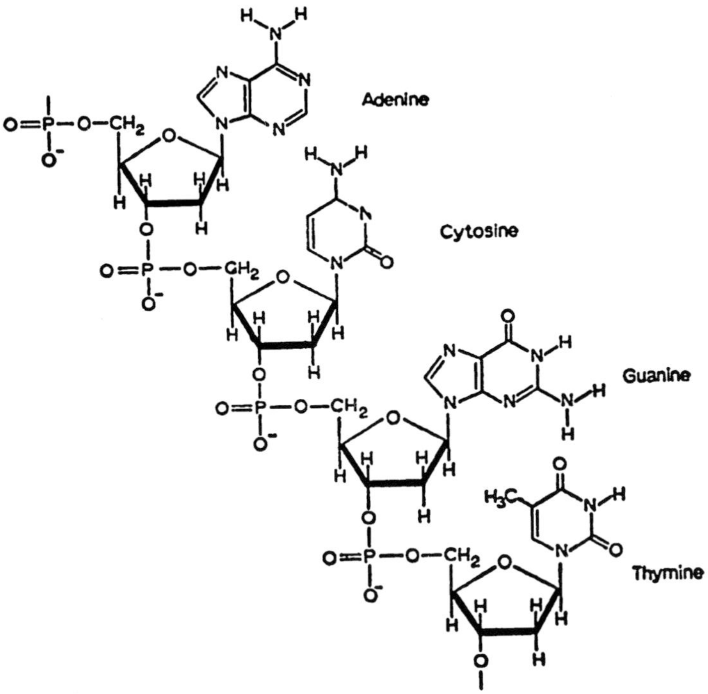

FIGURE 1.1 The chemical formula of a chain of a DNA molecule (D. M. Brown and A. R. Todd, *Journal of the Chemical Society*, 1952, 52. Reproduced by permission of the Royal Society of Chemistry). The backbone is made up of alternating sugar (2-deoxyribose) and phosphate groups. Each sugar has attached to it a side-group by a glycosidic linkage. The side groups consist of either a purine base (adenine or guanine) or a pyrimidine base (cytosine or thymine).

Note that the backbone has a directionality because the phosphate group is linked differently to the sugars on either side (to the 3′ carbon atom of one sugar and to the 5′ carbon atom of the other). A phosphate-sugar linked to a base is called a nucleotide. The DNA chain is synthesised from such nucleotides in the 5′–3′ direction.

The structure of DNA

Since the structure of DNA is so well known there is little point in keeping it to the end as a dénouement of this essay. Rather, it will help in what follows, if the structure and some of its implications are briefly described now. The basic chemical formula is shown in Figure 1.1.

The double helix (Figure 1.2) consists of two intertwined helical phosphate–sugar backbones, with the heterocyclic DNA bases projecting

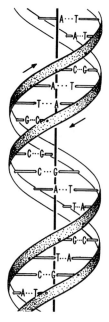

FIGURE 1.2 Schematic illustration of the DNA double helix as later sketched by Watson (*The Double Helix*, 1968).

inwards from each of the two strands. The two chains are antiparallel, running in opposite directions, and are related by a 2-fold axis of symmetry (dyad) perpendicular to the axis of the double helix. The bases are arranged in purine–pyrimidine pairs, adenine with thymine, guanine with cytosine, linked by hydrogen bonds (Figure 1.3), and these base pairs are stacked along the helix axis at a distance of 3.4 Å apart. The glycosidic bonds (the links between sugar and base) are related by the perpendicular dyad, so that they occur in identical orientations with respect to the helix axis. The two glycosidic bonds of a pair will not only be the same distance apart for all pairs, but can be fitted into the structure either way round. This feature allows all four bases to occur on both chains, and so *any* sequence of bases can fit into the double helix.

The two chains are said to bear a complementary relationship to each other. This means, as Crick and Watson spelt out in their second paper in *Nature*, that when the two chains come apart during replication of DNA, each can be used as a template to assemble a duplicate of its former partner (Figure 1.4). The crucial feature of the structure of DNA is not therefore the actual double helical form of the two phosphate–sugar chains – eye-catchingly iconic as it is (Plate I) – but the unique pairing of the bases projecting from each strand.

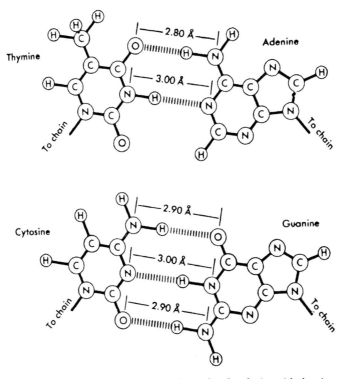

FIGURE 1.3 The pairing of bases by hydrogen bonds: adenine with thymine, guanine with cytosine (Crick and Watson (1954), *Proceedings of the Royal Society* A **223**, 1954, 80–96, who showed only two hydrogen bonds for the G–C pair, though tentatively suggesting a third, later confirmed by Pauling. Reproduced by permission of the Royal Society).

Structural research at King's College, London

In 1945, most biochemists, including sometimes Avery himself, had doubts whether something as simple as what DNA was thought to be – repeats of the four nucleotide bases – could be the genetic substance. More complex molecules like proteins – chromosomal proteins – were thought to be more likely candidates.

There were some who did believe in DNA, in particular, the 'phage group' in the USA led by Max Delbrück and Salvador Luria. The group, mostly geneticists, studied bacterial viruses, bacteriophages; they imagined, and later showed, a phage injecting its transforming principle, its DNA, its genes, into the host bacterium. A younger member of that group was James Watson, who

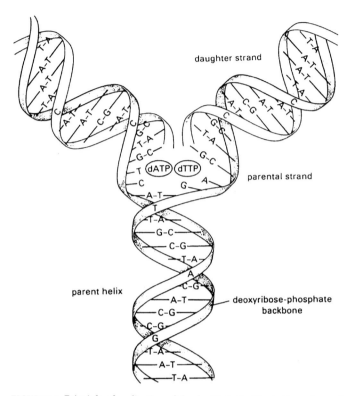

FIGURE 1.4 Principle of replication of the double helix. The helix unzips and each chain acts as a template for the synthesis of a complementary chain, thus creating two double helices, which are identical copies of the first.
'dATP' and 'dTTP' denote adenine and thymine nucleotide triphosphates, respectively, (energy-rich) monomers being incorporated at the next step in the growing chains.

in October 1950 went to Copenhagen to do postdoctoral research in nucleic acid chemistry, but was converted to a structural approach by hearing Maurice Wilkins speak at a Conference on Large Molecules in Naples in May 1951. Wilkins described his X-ray diffraction studies on fibres of DNA and showed diffraction patterns with much more detail than had been obtained by earlier workers Astbury and Bell in 1938. Moreover, they indicated a degree of crystallinity which raised the possibility of a molecular interpretation by X-ray analysis. Watson thereupon decided to go to a laboratory where he might learn X-ray diffraction techniques, and, failing to interest Wilkins, he eventually moved his fellowship to the MRC Unit in Cambridge headed by Max Perutz.

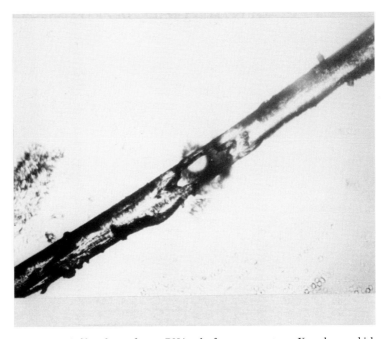

FIGURE 1.5 A fibre drawn from a DNA gel, after exposure to an X-ray beam which has punched a hole in it (M. H. F. Wilkins, W. E. Seeds and R. G. Gosling, *Nature* **167**, 1951, 759), reproduced from Gosling's Ph.D. thesis, King's College, London, 1954.

Here the structures of the proteins haemoglobin and myoglobin were being tackled. Watson arrived in September 1951 and met Francis Crick, who was working for his Ph.D. on haemoglobin, and found him like-minded about the importance of DNA. The rest is part of history, this history.

Wilkins was a senior member of the MRC Biophysics Unit at King's College, London, set up by (Sir) John Randall in 1946 after the Second World War to carry out 'an interdisciplinary attack on the secrets of chromosomes and their environment'. Wilkins worked to develop special microscopes, but having heard of the greatly improved methods devised by Rudolf Signer at Berne for extracting long unbroken molecules of DNA, he obtained some of the material and found a way of drawing uniform fibres from a viscous solution of DNA (Figure 1.5). Examination under polarised light showed them to be well-ordered, characteristic of long molecules oriented parallel to one another. He enlisted the help of a graduate student in the Unit, Raymond Gosling, who was studying ram sperm by X-ray diffraction. By keeping the fibres in a wet

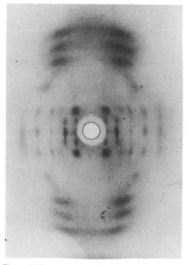

First multifibre specimen taken on the Raymax tube Unicam Camera, filled with hydrogen

FIGURE 1.6 Early X-ray diffraction patterns obtained by the King's College group, about 1950, suggestive of a helical structure (cf. A. R. Stokes, in *Genesis of a Discovery: DNA Structure*, ed S. Chomet, 1993).

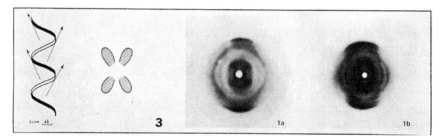

FIGURE 1.7 The first clear crystalline pattern from a DNA fibre, King's College, 1950, in what was later called the 'A form' (from R. G. Gosling in *Genesis of a Discovery: DNA Structure*, ed. S. Chomet, 1993). This was shown by Maurice Wilkins at the Naples conference attended by James Watson (courtesy, R. G. Gosling).

atmosphere, Wilkins and Gosling obtained the X-ray diffraction photographs he later showed at Naples and which so excited Jim Watson (Figure 1.6). Other early diffraction photographs of various specimens (Figure 1.7) showed hazy patterns, later understood to be indicating helical features (Plate II).

X-ray diffraction of crystals and fibres

X-ray crystallography provides a way of deducing the structure of a molecule by analysing the diffraction pattern produced when a beam of X-rays falls on

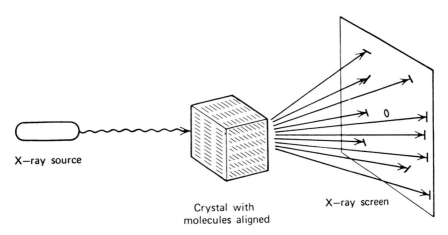

X-ray source

Crystal with
molecules aligned

X-ray screen

0

FIGURE 1.8 X-ray diffraction by a crystal containing a regular 3-D arrangement of molecules. The pattern of diffracted waves depends on the particular setting of the crystal relative to the incident X-ray beam. A full 3-D set of X-ray data is collected by rotating the crystal into different settings.

a crystal in which the molecules are regularly arranged in three dimensions (Figure 1.8). The pattern is nothing like a conventional photograph: it shows a set of spots of varying intensity and inferring the structure from the pattern is not a direct process. This is because each spot corresponds to a diffracted wave from molecules lying in a particular set of planes in the crystal. The molecular structure of the crystal could be reconstructed mathematically from a knowledge of the amplitudes and phases of the diffracted waves – 'amplitude' means strength of the wave (which is measurable from the spot intensity); and 'phase' means the positions of the peaks and troughs of the wave relative to the photographic plate, but the phase is lost in the recording. Hence arises the so-called 'phase problem' in X-ray crystallography which is that of developing methods for indirectly determining these lost phases. For small molecules, analytical methods have been developed, and for large molecules like proteins the problem was solved in 1953 by Max Perutz by the heavy atom isomorphous replacement method.

Fibrous macromolecules – polymers of small units, the monomers, regularly (or 'equivalently') arranged – present a further challenge in X-ray diffraction, since, in fibres, the long molecules, though roughly parallel to one another, are usually not all rotationally oriented relative to one another in a regular manner. The observed diffraction pattern then represents the rotational average of the patterns that would be given by different orientations. If the chemical structure of the monomer is known, as was, for example, the case with rubber (solved by C. W. Bunn in the late 1940s), then the polymer structure can be solved, by building models and comparing the calculated diffraction patterns with the observed ones. This is the model-building approach, which was used by Watson and Crick for DNA. The problem is that there is rotation about the single chemical bonds between monomers (and also usually within them), so other constraints must be used to fix how the monomers join head to tail. For a polypeptide chain, there are only two parameters, or degrees of freedom, to define the whole structure. Different combinations of these were used by Pauling in 1951 in predicting the α-helix and β-sheet of proteins. For a polymer of nucleotides, however, there are several more degrees of freedom (formally five, of different degrees of importance), and in the end these have to be fixed by fitting the model to the X-ray data; and the more detailed the latter is, the better the chance of succeeding. As described later, Watson and Crick used Franklin and Gosling's data. The final, extensive refinement, after 1953, of the DNA structure by Wilkins and his colleagues used their much-improved X-ray patterns.

Rosalind Franklin

In January 1951, the King's College group was strengthened by the arrival of Rosalind Franklin (Figure 1.9). She was a physical chemist, whose wartime work was on cokes and chars to see how they might be used more efficiently. She realised that to understand their internal structure it was necessary to use X-ray diffraction methods. In 1947 she moved to Paris where she was employed as Chercheur of the CNRS (Centre National de Recherche Scientifique). There she learned, and improved, X-ray diffraction techniques for dealing quantitatively with substances of limited internal order. These presented much more difficulty than the highly ordered crystals which X-ray crystallographers were using to solve the structures of small molecules. It is important to realise in what follows that, in Paris, Franklin gained no experience of such formal X-ray crystallography. There, however, she discovered the fundamental distinction between carbons that turned into graphite on heating and those that did not, results which later proved to be highly relevant for the development of carbon fibres.

The combination of these X-ray diffraction techniques and chemical preparatory skill attracted the attention of Randall, and Franklin was invited by him

FIGURE 1.9 Rosalind Franklin on holiday in Italy in the summer of 1950 (photo from V. Luzzati, Paris).

to bring her experience to King's College, London, for the purpose of studying the structure of DNA. Randall's purpose was clearly to put more professional effort into the DNA work begun by Wilkins and Gosling. Randall, however, left an unfortunate ambiguity about the respective positions of Wilkins and Franklin, which later led to dissension between them about the demarcation of the DNA research at King's. To this must be added the very different personalities of the two. A letter from Randall to Franklin in December 1950 (Figure 1.10) makes it clear that on 'the experimental X-ray effort...there will be at the moment only yourself and Gosling'. Wilkins did not know of this letter, and was away when Franklin arrived and Gosling was formally placed under her supervision.

Wilkins handed over the Signer DNA to them, and turned to an X-ray study of sperm where DNA is complexed with proteins. It should be remembered that at the time, no one, not even Watson and Crick, had imagined that the 3-D structure of DNA alone, important as that might turn out to be, would *by itself* indicate how the molecule replicated itself, and hence reveal 'the secret of life'. (In fact, Wilkins obtained good X-ray patterns, from oriented sperm heads of cuttlefish showing the helical B form of DNA).

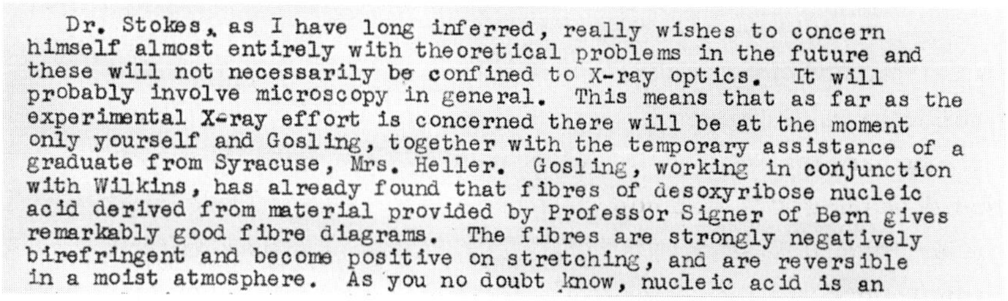

Dr. Stokes, as I have long inferred, really wishes to concern
himself almost entirely with theoretical problems in the future and
these will not necessarily be confined to X-ray optics. It will
probably involve microscopy in general. This means that as far as the
experimental X-ray effort is concerned there will be at the moment
only yourself and Gosling, together with the temporary assistance of a
graduate from Syracuse, Mrs. Heller. Gosling, working in conjunction
with Wilkins, has already found that fibres of desoxyribose nucleic
acid derived from material provided by Professor Signer of Bern gives
remarkably good fibre diagrams. The fibres are strongly negatively
birefringent and become positive on stretching, and are reversible
in a moist atmosphere. As you no doubt know, nucleic acid is an

FIGURE 1.10 Excerpt from a letter dated 4 December 1950 from Professor J. T. Randall, King's College, London, to Rosalind Franklin in Paris (Franklin papers, Churchill College Archives, Cambridge).

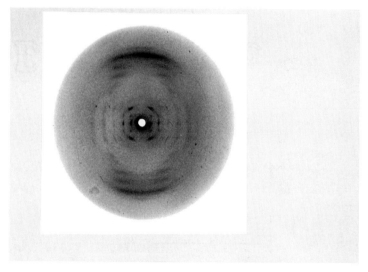

FIGURE 1.11 X-ray diffraction pattern of a DNA fibre, 1951, later understood to be a 'mixture' of the A and B forms of DNA (Gosling, Ph.D. thesis, London, 1954).

Franklin began a systematic study of DNA fibres and within the first year transformed the state of the field. By drawing thinner fibres she was able to enhance the alignment of the DNA molecules within the specimen, and these specimens, together with finer collimation of the X-ray beam generated from a microfocus X-ray tube which she and Gosling had assembled, produced sharper diffraction patterns (Figure 1.11). These, however, showed variable features, and it was not until Franklin made a systematic study of the fibres that the problem was solved.

The A and B forms of DNA

In a crucial advance, Franklin controlled the relative humidity in the camera chamber by using a series of saturated salt solutions and thus was able to regulate, and vary, the water content of the fibre specimens. In this way she showed that, depending on the humidity, two forms of the DNA molecule existed, which she later named A and B, and defined the conditions for the transition between them (Figure 1.12). The A form, which she first called 'crystalline', is found at and just below 75% relative humidity, and above that point there is an abrupt

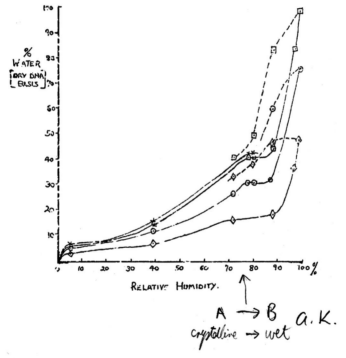

FIGURE 1.12 Franklin's measurement of water uptake by DNA fibres made by weighing them at different ambient relative humidities, the latter controlled by the use of saturated salt solutions (Franklin papers, Churchill College Archives, Cambridge; hitherto unpublished, described qualitatively in Franklin and Gosling, *Acta Crystallographica* **6**, 1953, 673). There is some hysteresis in that the curves on re-drying do not follow exactly the original wetting curves.

There is a marked transition at about 75% relative humidity, when the DNA structure changes abruptly from the 'crystalline' form (later called A) to the 'wet' form, B. Annotation in Franklin's notebook by the author of this essay Aaron Klug.

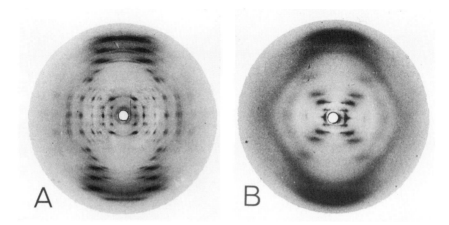

FIGURE 1.13 a and b X-ray diffraction patterns of the A and B forms of DNA
(Franklin and Gosling, *Acta Crystallographica* **6**, 1953, 673, Figures 1 and 4
respectively); the B form pattern is that reproduced in Franklin and Gosling, *Nature*
171, 25 April 1953, 740, also known as 'Photograph B51'.

transition to the B form, which she originally called 'wet'. Figure 1.13a and b
show the two X-ray patterns of A and B, respectively, and Figure 1.14 shows
the arrangement of molecules, whatever their internal structure might be, in
the two forms. The A form is crystalline in that the fibre is composed of small
crystallites in random orientations within it. In the B form fibre, it is the individual molecules themselves which, though parallel, are in random rotational
orientations, the molecules being separated by thin sheaths of water.

It became clear that all previous researchers had been working, unknowingly,
mostly with a mixture of the two forms, or at best with relatively poorly oriented
specimens of the A form, and, in retrospect, with occasionally hazy pictures of
the B form. It is easier to follow the rest of the story if one runs ahead to show
the first interpretations of the two forms (Figure 1.15).

The B form pattern illustrated in Figure 1.13b, is the superb and famous picture B51, which Franklin obtained later in May 1952, and which has achieved
iconic status. It was this picture that was shown by Wilkins to Watson in early
1953, and prompted the Cambridge pair into active model-building again, after
an earlier failure. Even less striking X-ray patterns, however, which Franklin
had obtained by September 1951,[1] showed clear evidence of a helical structure.

---

[1] R. E. Franklin and R. G. Gosling, *Acta Crystallographica* **6**, 1953, 673, Figure 2.

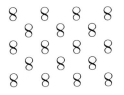

perfect crystalline array:
all molecules in same orientation

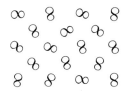

disordered array:
molecules in random orientations
(B form DNA fibre)

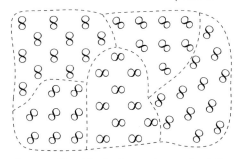

small crystalline blocks in random orientations
(A form DNA fibre)

FIGURE 1.14 The packing of DNA molecules in the A and B forms compared with those in a perfect crystal. The diagrams show schematic cross-sections of the arrangements.

The theory of diffraction by a helix had been worked out, at the behest of Wilkins, by Alex Stokes at King's (unpublished, 1951) and independently by Cochran, Crick and Vand the same year, published in early 1952. The characteristic feature of the pattern is the X-shaped pattern of streaks arranged in a set of layer lines, from which it can be deduced that the pitch of the helix is 34 Å.

A strong X-ray reflection lies on the meridian, corresponding to a spacing of 3.4 Å: this is produced by the regular stacking of the bases on top of each other, (as indeed, had been proposed by Astbury and Bell in 1938 who could see this strong reflection in their hazy X-ray patterns). Since the helix pitch is 34 Å, this means that the helix, whatever it is in detail, repeats after 10 (= 34/3.4) units per turn. This photograph is particularly striking in that it shows not only the X-shaped pattern of streaks in the centre, but also secondary fans emanating from the two 3.4 Å meridional reflections, top and bottom, and

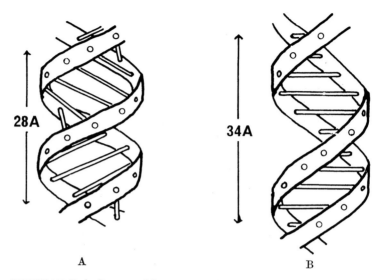

FIGURE 1.15 Early diagrams of the structures of the A and B forms of DNA (G. B. Sutherland and M. Tsuboi, *Proceedings of the Royal Society* A **239**, 1957, 446; the A form after Wilkins *et al. Nature* **172**, October 1953, 759).

running obliquely to the equator. These are characteristic of a discontinuous helix (Figure 1.16), as is to be expected from a phosphate–sugar chain.

In the A form, the repeat of the structure is 28 Å compared with 34 Å in the B, consistent with a macroscopic shrinkage of 25% in the lengths of the fibres. The A form does not show the characteristic X, but there is a gap on the meridian of the photograph, consistent with a helical structure, as Franklin recognised (see below, on her Colloquium in November 1951). However, this might also be accounted for by sheets or rods packed at an inclined angle to the fibre axis and this latter interpretation was for a time later pursued by Franklin after she took a wrong turning in her attempts to analyse the A form. The A form, as later recognised by Watson, is a somewhat more tightly wound form of the B double helix, in which the bases change their tilt obliquely to the fibre axis (Figure 1.15), thus obscuring the characteristic X-shaped fan of reflections expected from a simple helical structure.

For, despite her discovery of the simpler B pattern, Franklin at first directed her attention mostly to the A form. Here the molecules themselves are not in random rotational orientations, as in B, but packed regularly in a crystal lattice (Figure 1.14). The X-ray pattern thus shows sharp 'spots' and

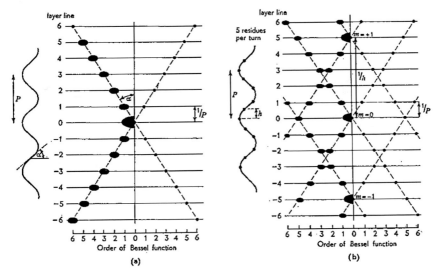

FIGURE 1.16 Diffraction pattern given by a discontinuous helix made of discrete units (right), compared with that given by a continuous helix (left, cf. Plate II). In this example, there are 5 units per turn of the helix, giving rise to a meridional reflection (i.e. on the axis) on the fifth layer line. There is a subsidiary X-shaped fan, emanating from the meridional reflection, as well as the main fan emanating from the centre of the pattern. This can be seen in Franklin's X-ray pattern of the B form (Figure 1.13b).

offered the possibility of an objective crystallographic analysis because of the greater wealth and precision of the diffraction data available. In retrospect this was a misjudgement, but it was a reasonable decision at the time, because, if correctly interpreted, the A pattern would yield more precise information about the DNA molecule. She decided to use what is called 'Patterson function analysis' on the X-ray data she had measured on the A patterns, and, as Gosling said later, to let the data speak for themselves. This Patterson method is an indirect method, which had been used at higher resolution to solve the structures of small molecules, but never for such large unit cells. A map of the Patterson function, directly calculable from the X-ray intensities, shows only vectors between elements of the structure (atoms, or groups of atoms), not the structure itself. The Patterson map was not successfully interpreted before the Watson–Crick model, but it was afterwards,[2] in terms of two helical chains (Figure 1.25, below).

[2] R. E. Franklin and R. G. Gosling, *Nature* 72, 1953, 156.

Franklin's colloquium, November 1951: Watson and Crick's first model

In November 1951, Franklin, before embarking on the Patterson analysis, held a colloquium which Watson, now at Cambridge, attended. As his book *The Double Helix* makes clear, there was much contact on and off between Wilkins and Crick, who were friends, and this led to several visits by Watson to King's.

The draft of Franklin's colloquium and her accompanying notes survive in the Archives of Churchill College, Cambridge. She describes the two forms as 'crystalline' (2) (later A) and 'wet' (3) (later B) as well as a (very) dry form (1), which is not easily re-wetted (hysteresis). She gives the crystal parameters, and the lattice symmetry (monoclinic space group C2), and also the density of A, from which she deduced that there were two, three or, less likely, four chains of DNA per lattice point. She notes the pseudo-hexagonal packing, which implies that the molecules have an approximately cylindrical shape with a diameter of about 20 Å. Her notes read: 'Evidence for spiral structure. Straight chain untwisted highly improbable. Absence of reflections on meridian in xtalline form suggests spiral structure...Nucleotides in equivalent positions occur only at intervals of 27 Å [corresponding to] the length of turn of the spiral.'

On the basis of the above, Franklin put forward her view that the molecular structure in the A form was likely to be a helical bundle of 2 or 3 chains, with the phosphate groups on the outside. The bundles are separated by weak links produced by sodium ions and water molecules (Figure 1.17). At the higher humidity of the B form, a water sheath disrupts the relationship between neighbouring helical bundles, and only the parallelism of their axes is preserved. (The same conclusions are found in Franklin's Fellowship Report for the year ending 1951.) Watson (and others) have stated in reminiscences that Franklin did not mention the B form, but her draft is quite explicit about the helical bundle being preserved in the transition from A to B. Indeed, her notes read 'Helical structure in 3 [wet] can't be the same as in 2 [crystalline] because of large increase in length.'

Watson took the news – as little, or as much, as he understood of it – back to Crick in Cambridge, and, now with some structural information to hand, they decided to build a model. They had urged this approach on the King's group, but, receiving no response, now felt justified in attempting this themselves. The King's group was invited to see the result – a model built in a week. The

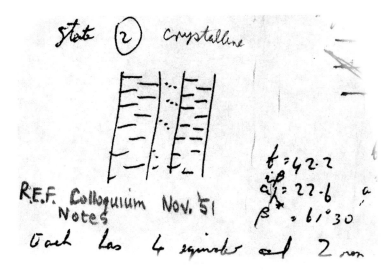

FIGURE 1.17 Diagram from Franklin's notes for the colloquium she gave at King's College in November 1951, annotated by Aaron Klug. The DNA molecules in the A form are represented as helical bundles of two, or three, chains (here two), with the bases on the inside, the phosphates on the outside, and the individual molecules associated in the fibre through water and ionic links (dotted lines). Each molecule has 6 near neighbours, 4 equivalently related and two others approximately related (Franklin papers, Churchill College Archives).

model was of three helical chains with the phosphates on the inside, neutralised by cations, with the bases pointing outwards. Franklin asked where was the water, and received the reply that there was none. It turned out that Watson, not understanding the relationship between a unit cell of a crystal and the asymmetric unit, had conveyed the wrong water content. After this debacle, Sir Lawrence Bragg, the head of the Cavendish Laboratory, firmly vetoed any further work on DNA at the MRC Unit in Cambridge. In future it would be done solely at the Unit at King's College.

Non-helical DNA?

Franklin pressed ahead with the Patterson analysis of the A Form. There is no question that all along she held the view that the B form was helical (Figure 1.18), but could not see a way to solve it except by model-building, a path she was reluctant to follow. Although she knew of Pauling's success in 1951 in predicting,

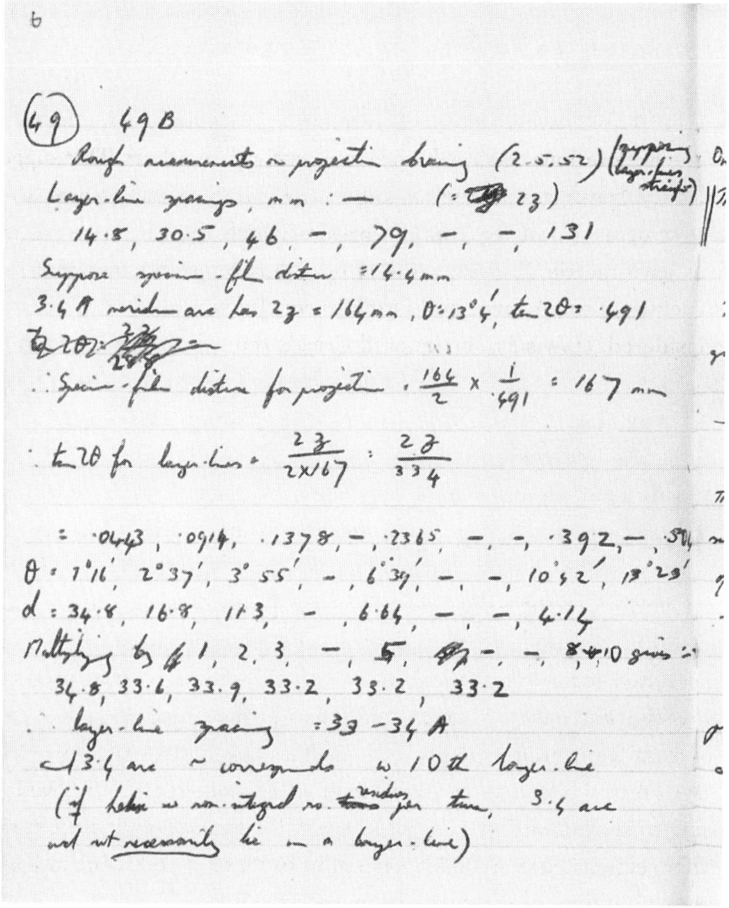
b

(49) 49B

Rough measurements on projection drawing (2.5.52) (gappings layer-lines strips)

Layer-line spacings, mm: — (÷ 23)

14.8, 30.5, 46, —, 79, —, —, 131

Suppose specimen-film distance = 16.4 mm

3.4 Å meridian arc has 2z = 164 mm, θ = 13°4', tan 2θ = .491

∴ Specimen-film distance for projection = 164/2 × 1/.491 = 167 mm

∴ tan 2θ for layer-lines = 2z/(2×167) = 2z/334

= .0443, .0914, .1378, —, .2365, —, —, .392, —, .39

θ = 1°16', 2°37', 3°55', —, 6°39', —, —, 10°52', 13°29'

d = 34.8, 16.8, 11.3, —, 6.64, —, —, 4.14

Multiplying by 1, 2, 3, —, 5, —, —, 8, 10 gives:

34.8, 33.6, 33.9, 33.2, 33.2, 33.2

∴ layer-line spacing = 33–34 Å

∴ 3.4 arc ~ corresponds with 10th layer-line

(if helix has non-integral no. residues per turn, 3.4 arc need not necessarily lie on a layer-line)

FIGURE 1.18 Excerpt from Franklin's notebook, May 1952, showing an analysis of the X-ray pattern 49B (a precursor of the famous 51B picture, Figure 1.13b above) in terms of helical diffraction theory (Franklin papers). This was done at a time when she was questioning whether the A form was helical.

by model-building, the α-helical and β-sheet configurations of the polypeptide chains of proteins, she equally well knew of the contemporary failure of Bragg, Kendrew and Perutz on the same problem – the 'greatest fiasco of my scientific life', Bragg later called it. This last episode of Watson and Crick would only have confirmed her decision to avoid model-building and rather to try an analytic crystallographic approach on the A form. She well knew, however, that the DNA structure would have to account for both the A and B forms.

However, an unfortunate mechanical accident in one of the specimens led Franklin to take a wrong turning. In the spring of 1952, one DNA fibre gave an X-ray pattern showing strong 'double orientation', that is, the 3-D crystallites in the A form were not all in random orientation about the fibre axis, but some orientations occurred more frequently than others. This suggested to her that the symmetry of the crystallite was far from cylindrical, so that *either* the structural unit itself *or* the way in which the units are packed together was asymmetric. If correct, the latter interpretation would rule out a helical structure in the A form, and Franklin concluded that this possibility had to be considered. Unwisely, she ignored Crick's remark to her, made in a tea-queue at a meeting, that the double orientation was an accident to be dismissed. It is this view of hers which gave rise to her supposed 'anti-helical' stance, but for her it was a *question* which had to be answered.

In fact she seems to have persuaded Wilkins, even though relations were strained between them, to the same view. Thus, ironically, while Franklin does not mention this in a Report written in late 1952 for an MRC Sub-Committee on the work of the King's Unit, Wilkins writes: 'The crystalline material gives an X-ray picture with considerable elements of simplicity which could be accounted for by the helical ideas, but three dimensional data show apparently that the basic physical explanation of the simplicity of the picture lies in some quite different and, *a priori*, much less likely structural characteristic. The 20 Å units, while roughly round in cross-section, appear to have highly asymmetric internal structure.' This is the MRC Sub-Committee Report, of which other sections gave crucial information to Watson and Crick in February 1953 for the building of their correct model of DNA.

This misjudgement on Franklin's part did, to some extent, influence her attempts to interpret the Patterson map of the A form. She sought an explanation in terms of rods or sheets, or a 'figure of eight', which naturally failed. She was apparently thinking of the A form as an unwound version of the helices in the B state (rather, I imagine, like the β-sheet structure is to the α-helix in proteins and polypeptides).

One correct result which emerged in January 1953, from her application of the so-called 'superposition method' to the map, was that the A form contained two chains, and that they ran in opposite directions (Figure 1.19). Had she been a crystallographer, and understood the meaning of the crystal symmetry, C2 face centred monoclinic, which she herself had established much earlier, she

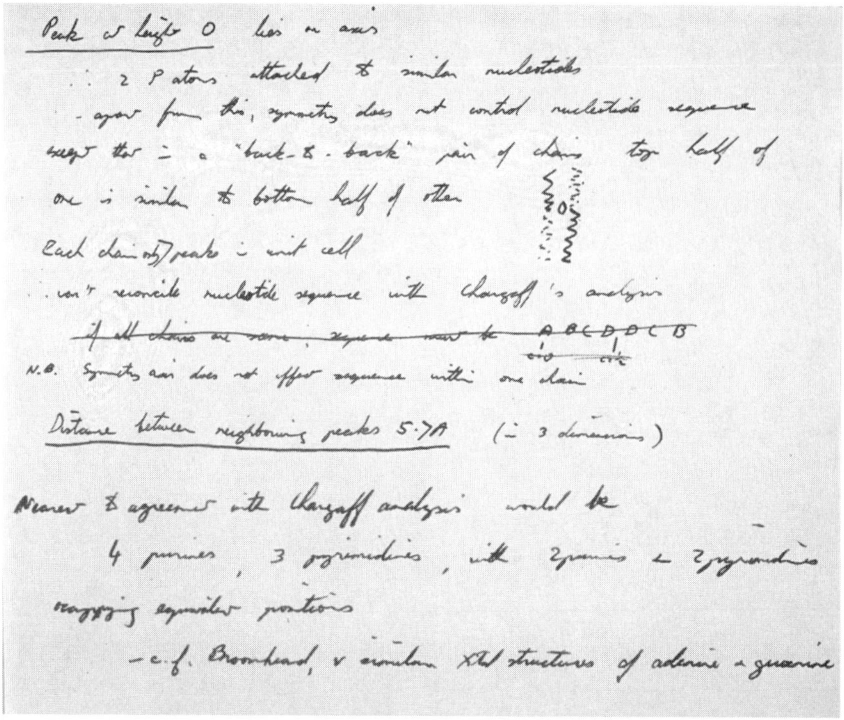
Peak at height 0 lies on axis
∴ 2 P atoms attached to similar nucleotides
- apart from this, symmetry does not control nucleotide sequence
except that in a 'back-to-back' pair of chains top half of
one is similar to bottom half of other
Each chain only 7 peaks in unit cell
∴ can't reconcile nucleotide sequence with Chargaff's analysis
~~if all chains are same, sequence must be~~ A B C D D C B
N.B. Symmetry axis does not affect sequence within one chain
Distance between neighbouring peaks 5.7A (in 3 dimensions)
Nearest to agreement with Chargaff analysis would be
4 purines, 3 pyrimidines, with 2 purines & 2 pyrimidines
occupying equivalent positions
– c.f. Broomhead, & similar Xtal structures of adenine & guanine

FIGURE 1.19 Excerpt from Franklin's notebooks on 19 January 1953 (Franklin papers). Franklin deduces by Patterson superposition analysis that the A form contains two chains related by a 2-fold axis of symmetry (the oval symbol). Also noted are Chargaff's base ratios, and Broomhead's crystal structures of adenine and guanine, which showed the correct tautomeric forms.

could have deduced this result at once. Of all the protagonists in the story, only Crick understood this and, moreover, this was the space group symmetry of the ox-haemoglobin crystals which he was studying for his Ph.D. under Perutz. It meant that, if the A structure was helical, it would consist of 2 chains, or strands, running in opposite directions, related by a 2-fold axis of symmetry perpendicular to the fibre axis, and hence to the pair of chains. (Franklin hardly ever reminisced about DNA in the years I worked with her at Birkbeck College, but she once said that she could have kicked herself for missing the implications of the C2 symmetry.)

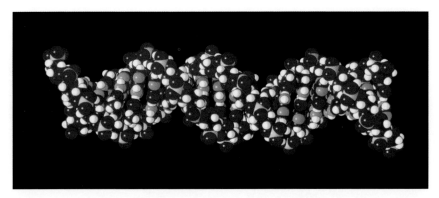

CHAPTER I PLATE I Space-filling atomic model of the DNA double helix. Colouring: phosphorus yellow; oxygen red; carbon dark blue; nitrogen light blue; hydrogen white.

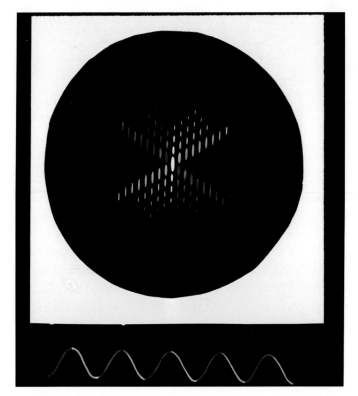

CHAPTER I PLATE II Diffraction pattern produced by a continuous helix: an optical analogue (from K. C. Holmes and D. M. Blow. 'The use of X-ray diffraction in the study of protein and nucleic acid structure', *Interscience*, 1966). Note the X-shaped fan of reflections emanating from the origin (at the centre).

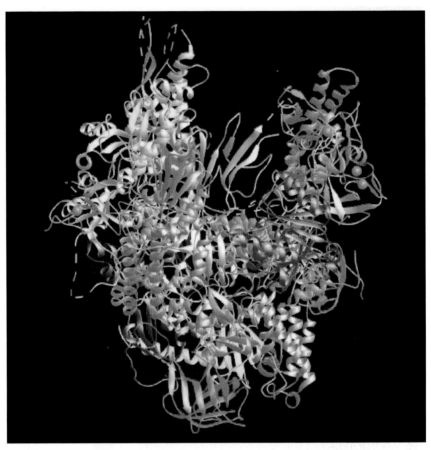

CHAPTER I PLATE IIIA Structure of RNA polymerase II, the central enzyme of DNA transcription. An end-on view to show the cleft (at the top) for locating the DNA double helix to be transcribed (Cramer *et al.*, *Science* **292**, 2001, 1863; courtesy of Roger Kornberg).

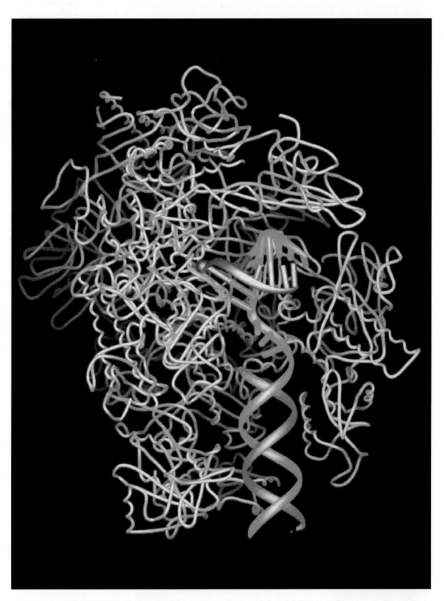

CHAPTER I PLATE IIIB Cut-away side view of a complex of RNA polymerase II, with a DNA double helix (blue) trapped in the act of transcription. The newly transcribed RNA (red) will exit from the top left-hand corner (Gnatt *et al.*, *Science* **292**, 2001, 1876; courtesy of Roger Kornberg).

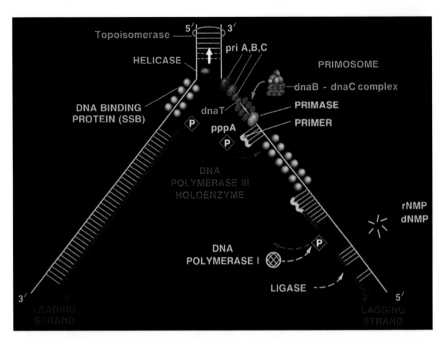

CHAPTER I PLATE IVA DNA replication. Scheme of DNA replication (courtesy, Arthur Kornberg). The DNA double helix (top) is cut at the replication fork by a topoisomerase enzyme, and unwound by a helicase, the separated strands being coated with single strand DNA binding protein (SSB). The leading strand (left) is copied into RNA (red) in straightforward way by the enzyme DNA polymerase III (the 'locomotive'). Since nucleic acids can be synthesised only in the 5′–3′ direction, the lagging strand (right) is synthesised by an elaborate mechanism, using RNA intermediates, from short DNA sequences, by DNA polymerase I (the 'sewing machine'). These are then linked covalently together by ligases.

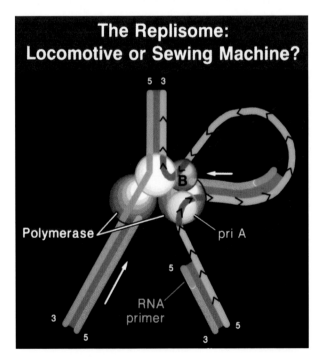

CHAPTER I PLATE IVB A schematic view of the combined replication machinery (courtesy, Arthur Kornberg). It is believed that the two DNA polymerases and the primase are grouped together at the replication fork, and move as a whole along the DNA, turning the parent double helix into two daughter double helices.

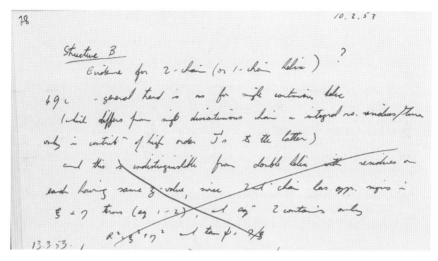

FIGURE 1.20 Excerpt from Franklin's notebooks on 10 February 1953 (Franklin papers). Franklin returns to the question of the number of chains in the B form.

Franklin's Patterson analysis ran into an impasse, and, in February, she turned to her B form, the X-ray pattern which was clearly characteristic of some kind of helical structure (Figure 1.20). Her notebooks show her shuttling back and forth between the two forms. She had by now abandoned her attempts to interpret the A form in non-helical terms. On 23 February (Figure 1.21) she writes 'If single-strand helix as above is basis of structure B, then Structure A is probably similar, with P-P distance along fibre axis < 3.4 Å, probably 2–2.5 Å.' On 24 February (Figure 1.22) she is at last making the correct connection between the A and B forms – both have 2 chains. Of course, she had no idea that, at that very time in Cambridge, Crick and Watson were now back to model-building of DNA. Nor were they aware of what Franklin had been doing – no wonder that Watson wrote in *The Double Helix* that Franklin's instant acceptance on first seeing their model amazed him. He had then no idea how close she had come to it.

By March 1953, using helical diffraction theory, Franklin had carried the quantitative analysis of her B form patterns to the point where the paths of the backbone chains were determined. She had moved to Birkbeck College to J. D. Bernal's Department of Physics on 14 March and there she wrote up

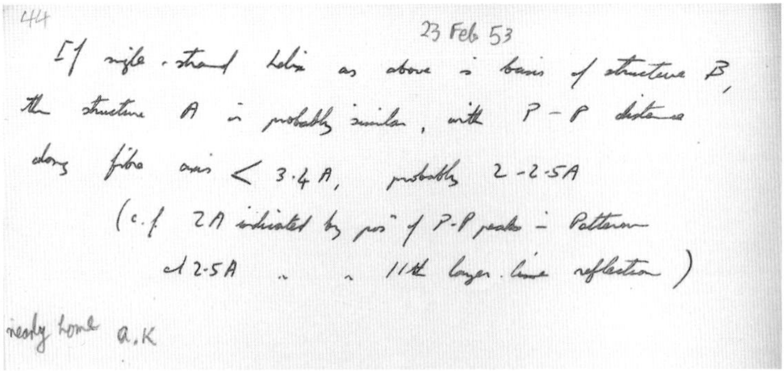

FIGURE 1.21 Excerpt from Franklin's notebooks on 23 February 1953 (Franklin papers). Annotation by Aaron Klug. Franklin begins to consider a helical structure for the A form.

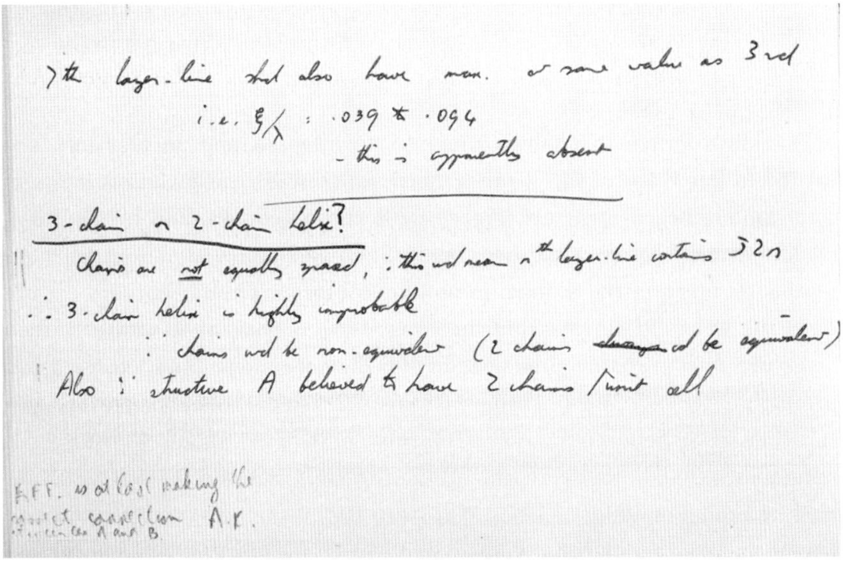

FIGURE 1.22 On 24 February 1953, Franklin comes down in favour of two chains for the B form, making a connection with the A form. Annotation by Aaron Klug.

her work in a typescript (on Birkbeck's Department of Physics' typewriter!) dated 17 March, that is, one day before the manuscript of Watson and Crick's structure, prepared for *Nature*, reached King's. Franklin's draft (Figure 1.23) contains all the essentials of her later paper (with Gosling) in *Nature* in April,

ROUGH DRAFT

A NOTE ON MOLECULAR CONFIGURATION IN SODIUM THYMONUCLEATE

Rosalind E. Franklin and R. G. Gosling

17/3/53.

Sodium thymonucleate fibres give two distinct types of X-ray diagram. The first, corresponding to a crystalline form obtained at about 75% relative humidity, has been des-cribed in detail elsewhere ( ). At high humidities a new structure, showing a lower degree of order appears, and persists over a wide range of ambient humidity and water content. The water content of the fibres, which are crys-talline at lower humidities, may vary from about 50% to several hundred per cent. of the dry weight in this structure. Other fibres which do not give crystalline structure at all, show this less ordered structure at much lower humidities. The diagram of this structure, which we have called structure B, shows in striking manner the features characteristic of helical structures ( ). Although this cannot be taken as proof that the structure is helical, other considerations make the existence of a helical structure highly probable.

FIGURE 1.23 Unpublished typescript dated 17 March 1953, which is the precursor of Franklin and Gosling's paper in *Nature*, 25 April 1953 (Churchill College Archives; see Klug, *Nature* **248**, 1974, 787).

which, together with one by Wilkins, Stokes and Herbert Wilson, accompanied Crick and Watson's paper announcing their model for the structure of DNA.

In Franklin's draft, it is deduced that the phosphate groups of the backbone lie, as she had long thought, on the *outside* of the two co-axial helical strands whose geometrical configuration is specified, with the bases arranged on the inside. The two strands are separated by 13 Å (three-eighths of the helix pitch in the axial direction). Her notebooks show that she had already formed the notion of the interchangeability of the two purine bases with each other, and

FIGURE 1.24 Francis Crick and James Watson with their model of the DNA double helix, 1953 (photo by Anthony Barrington Brown/Science Photo Library).

also of the two pyrimidines. She also knew the correct tautomeric forms of at least three of the four bases, and was aware of Chargaff's base ratios (see below). The step from interchangeability to the specific base pairing postulated by Crick and Watson is a large one, but there is little doubt that Franklin was poised to make it. The draft shows that, although she had earlier concluded that the A form contained two chains or strands which ran in opposite directions (Figure 1.19), she had not yet grasped that the two chains in B would also have to be antiparallel.

What would have happened if Watson and Crick had not intervened with their great burst of insight (Figure 1.24), and Franklin had been left to her own resources? It is a moot point whether she was one and a half or two steps behind, and how long it would have taken her to take them. Crick and I have discussed this several times. We agree she would have solved the structure, but the results would have come out gradually, not as a thunderbolt, in a short paper in *Nature*.

Pauling's entry into the field

I must now, however, move the story back to Cambridge in early 1953, when Crick and Watson re-entered the scene, having received permission from Lawrence Bragg to do so. News had reached them that Linus Pauling, the greatest chemist of the day, had a structure for DNA and that a manuscript was on its way. Here an old rivalry asserted itself: I was told the story by R. W. James, my Professor at the University of Cape Town, who had been a close collaborator of Bragg at Manchester. In the mid 1930s Pauling used some of Bragg's unpublished results on silicate structures, gleaned during a visit to Manchester, together with his own data, not revealed to Bragg, to formulate his famous rules which explained how the wide variety of silicate minerals are formed. Then later there was the chagrin Bragg felt, as described above, at having missed Pauling's α-helix.

Pauling's manuscript arrived at the Cavendish Laboratory in the last week of January, and it was immediately obvious he had made a crucial chemical mistake in postulating a 3-chain structure with a central phosphate–sugar backbone. It resembled Crick and Watson's wrong model of November 1951, but the phosphate groups were not ionised, so there was nothing to hold the chains together. It was chemically impossible, but no doubt Pauling would return, or so Watson argued to Bragg. (I doubt this – Pauling was a man with great insight, but not a magician who could manage without data.)

Watson and Crick's structure

The fact that Pauling was now in competition made for a race, and since the King's group seemed to be divided and making no progress, Bragg was persuaded to unleash Crick and Watson from his earlier ban. The urgency was conveyed to him by Watson, who, two days earlier, had visited King's to give a copy of the Pauling manuscript to Wilkins. In the course of that visit Wilkins showed him Franklin's striking May 1952 X-ray picture of the B form, with its clear helical features. This made a profound impression on Watson, since one could also immediately count the number of layer lines leading to the 3.4 Å meridional reflection. He recounted this to Crick, along with the other parameters necessary to build a B form model: the repeat distance of 34 Å, indicating 10 units per helical turn, a helix slope of 40°, the diameter of about 20 Å of the molecule, and they also remembered Franklin's arguments for the backbones being on the outside of the molecule.

The rest of the story is told in Watson's vivid account in his book *The Double Helix*, but he glosses over the details of the information in the MRC Sub-Committee Report on the work at King's which came into their hands in the second week of February. This was given them by Max Perutz, a member of that Sub-Committee. Excerpts from the Report itself and a discussion on its degree of confidentiality by Perutz, Wilkins and Watson were later published in *Science*.[3] The Report confirmed much that they already knew, but the key fact, as I mentioned earlier, was the space group symmetry C2 of the A form. Franklin had disclosed this in her colloquium in November 1951, but Watson would not have understood it. Crick had heard that the crystal was monoclinic, which implied a 2-fold axis of symmetry (a dyad), but this could have been parallel or perpendicular to the fibre axis. C2 required it to be perpendicular to the fibre axis.

Watson had begun the building of 2-chain helical models with the chains running in the same direction. Each chain would repeat after 20 nucleotides. This would fit the best estimate of the number of nucleotides per lattice point (16 to 24, deduced from Franklin's density measurements on the A form) which could be reconciled with the 10-fold repeat. The chain had a helical rotation angle of 18° (= 360°/20) between nucleotides, which brought successive sugars close together and was difficult to build. The C2 symmetry, however, told Crick that there were indeed 2 chains, that the chains ran in opposite directions and that the helical repeat of 10 units per turn referred to one chain, that is, to each of the chains. Crick therefore changed the rotation angle to 36° and Watson found the chain easier to build. This was a crucial step in getting the backbone structure right.

The formal account by Crick and Watson in the *Proceedings of the Royal Society*,[4] which details their cogent reasoning in arriving at the double helix, does not mention their knowledge of the C2 symmetry which they had

[3] *Science* **164**, 1969, 1537–9.
[4] *Proceedings of the Royal Society* **223**, 1954, 80–90.

obtained from the MRC Report. (Neither does Watson's book.) It acknowledges information received from Wilkins and Franklin only in general terms: 'We are most heavily indebted in this respect to the King's College group, and we wish to point out that without this data the formulation of our structure would have been most unlikely, if not impossible'. Presumably there would have been some embarrassment about mentioning the source of their knowledge of the C2 symmetry. It should nevertheless have been stated.

The next step facing Watson and Crick was to fit the bases stacked at 3.4 Å above each other into the middle of the double helix. The bases are linked by glycosidic bonds to the sugars of the backbones. There was room for two bases in each stack and the glycosidic bonds had to be related by the perpendicular dyad. Watson tried different ways of making such pairs, connected by hydrogen bonds, initially pairing like with like, thus, adenine with adenine, and so on. In the last week of February, it was, however, pointed out to Watson by Jerry Donohue, who shared an office with him and Crick – another chance event – that he was using the incorrect chemical formulae (tautomeric forms) for the four bases. When Watson changed these he found he could fit in adenine–thymine as a pair, and guanine–cytosine as a pair. As described earlier, the geometry of each pair was almost identical! Remarkably, this pairing gave an explanation of the 'rules' (rather, 'ratios') found some years earlier by Erwin Chargaff on the chemical composition of the bases in DNA, namely, that the amount of adenine in any DNA sample equals that of thymine, and similarly for guanine and cytosine. Chargaff's ratios thus automatically arose as a consequence of a double helix structure of DNA and Watson's base pairing scheme. The structure of DNA was solved!

On 28 February 1953, Crick 'winged' in to the Eagle pub, close to the Cavendish Laboratory, where lunch could be had for 1s 9d, and declared to anyone who cared to listen that, in the Cavendish, Watson and he had discovered 'the secret of life'. It took a week longer to adjust the model using aluminium models of the bases (impatiently, Watson had earlier gone ahead with rather inaccurate cardboard cut-outs), and a few more days to pass the news to King's. Wilkins came to see the model in mid March, and Franklin later at the end of the month. Her 'instant acceptance' amazed Watson, but then he had no idea how far she had got towards it, having only heard of her supposed 'anti-helical' stance.

There was agreement between King's and Cambridge to publish separately, and three papers appeared on 25 April 1953, grouped together under the overall title 'Molecular structure of nucleic acids'. Watson and Crick's paper contained what appeared to be the famous throw-away sentence: 'It has not escaped our notice that the specific [base] pairing we have postulated immediately suggests a possible copying mechanism for the genetic material.' Crick explained later that they were not being coy, but there was a worry on Watson's part that the structure might be wrong:[5] when they sent the first draft of the paper to King's, they had not yet seen their papers and had little idea of how strongly the King's X-ray evidence supported their structure. After seeing it they wrote their second *Nature* paper of 30 May entitled 'Genetical implications of the structure of deoxyribonucleic acid'.

Proving the model

The first analytical demonstration of the general correctness of the Watson–Crick model came in July 1953 from Franklin and Gosling (Figure 1.25). They showed that their Patterson function map of the A form could be fitted by a helical structure with two chains, although the distance deduced between the two chains is only approximate, the X-ray data to a resolution of only 3 Å being no more than sufficient for the main purpose. Franklin left King's College for Birkbeck College in March 1953, where she took up the problem of the structure of tobacco mosaic virus.

The task of rigorously testing the model against X-ray diffraction data required more accurate intensity data from better-oriented fibre specimens and this was undertaken by Wilkins and the King's College group. It took them about seven years to carry this out. They obtained much-improved diffraction patterns from several different DNA sources (Figures 1.26 and 1.27), built higher-resolution X-ray cameras, introduced computers to make the calculations and new analytic methods for refining models to fit X-ray fibre diffraction. During that time there were several objections by crystallographers to the DNA model. The correctness of the base pairing scheme was questioned by Donohue (there are 24 different ways of pairing bases), and a crystal structure of an isolated G-C base pair determined by Hoogsteen showed a different

---

[5] *Nature* **248**, 26 April 1974, 766–9.

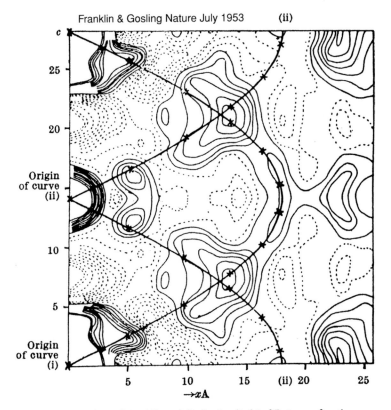

FIGURE 1.25 Analysis of Franklin and Gosling's cylindrical Patterson function map of the A form in terms of a double helix (*Nature* **172**, 25 July 1953, 156). The curves denote the 'self Pattersons' of the two helical chains, separated by half the helical pitch. The fit is improved if the 'cross-Patterson' between the two chains is included (D. L. D. Caspar, private communication, 1968).

pairing from that necessary to fit into the double helix. These and other objections were finally answered by the rigorous analysis at King's, although other models appeared occasionally through the 1960s and 1970s. Indeed, it could be said that the formal crystallographic proof of the double helix and the base pairing did not come until 1979, when Drew and Dickerson solved the structure of a dodecameric DNA oligonucleotide of defined sequence, by using the totally objective heavy atom method.[6]

[6] *Proceedings of the National Academy of Sciences of the USA* **78**, 1981, 2178–83.

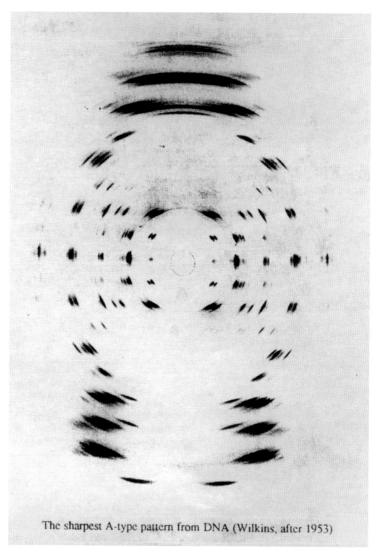

The sharpest A-type pattern from DNA (Wilkins, after 1953)

FIGURE 1.26 The sharpest, high-resolution A type diffraction pattern obtained by Wilkins and the King's group, post 1953. (courtesy Maurice Wilkins, in *Genesis of a Discovery*, ed. S. Chomet, 1993).

One of the sharpest crystalline B-type patterns (Wilkins, after 1953)

FIGURE 1.27 High-resolution, sharp, X-ray diffraction pattern of a *crystalline* B form, post 1953, obtained from a lithium salt of DNA (Wilkins, in *Genesis of a Discovery*, ed. S. Chomet, 1993). The outermost spot corresponds to a spacing of 1.7 Å, the second order of the spacing of 3.4 Å between the bases.

The reception of the double helix

It should be remembered that, in 1953, X-ray diffraction and crystallography of large biological molecules was still in its infancy and regarded as an exotic pursuit; the first protein structures of myoglobin and haemoglobin were not solved (at low resolution) until 1957 and 1959 respectively.

The double helix model was well received by geneticists and the phage group when Watson described it at the Cold Spring Harbor meeting in the summer of 1953, but there were doubts about the correctness, and indeed relevance, of the model on the part of biochemists, who, on the whole, still thought of proteins as the genetic material. The best biochemical proof that the structure was correct eventually came from Arthur Kornberg, in the course of his studies on the enzymatic synthesis of DNA. If the 'hypothetical' dyadic structure of DNA with 2 antiparallel chains (Figure 1.2) were correct, then there must also be relationships between pairs of dinucleotides, further to Chargaff's rules for individual bases. Thus the number of $\overrightarrow{AG}$ dinucleotides should equal the number of $\overrightarrow{CT}$ dinucleotides, the number of $\overrightarrow{TG}$ be equal to that of $\overrightarrow{CA}$, and so on. Kornberg and his colleagues measured the frequencies of dinucleotides in a variety of DNAs.[7] The prediction was proved correct, in a most elegant way.

Furthermore, the structure of the double helix, as emphasised by Todd, was still only a discovery in chemistry, and, even if correct, the biological implications for replication ('semi-conservative replication') as postulated by Crick and Watson, persuasive as they were, did not necessarily follow. The proof came in 1958 from 'the most beautiful experiment in biology' (the title of a recent book) by Meselson and Stahl.[8] This demonstrated unequivocally that the complementary strands of a DNA molecule separate from one another, and that each strand then serves as the template for the synthesis of a complementary strand, duplicating its former partner, and so producing two DNA double helices (Figure 1.28).

Biochemists and biologists generally also began to understand that the Watson–Crick base pairing allowed *any* 'random' or irregular sequence of the four bases of DNA to be accommodated within the double helix, and so it was possible for DNA to act as the carrier of genetic information based on a four-letter code A, G, C, T. In 1962, the Nobel Prize for Physiology and Medicine was awarded to Crick, Watson and Wilkins. Rosalind Franklin had died in 1958,

---

[7] Josse *et al.*, *Journal of Biological Chemistry* **236**, 1961, 864–75.
[8] *Proceedings of the National Academy of Sciences of the USA* **44**, 1958, 671–5.

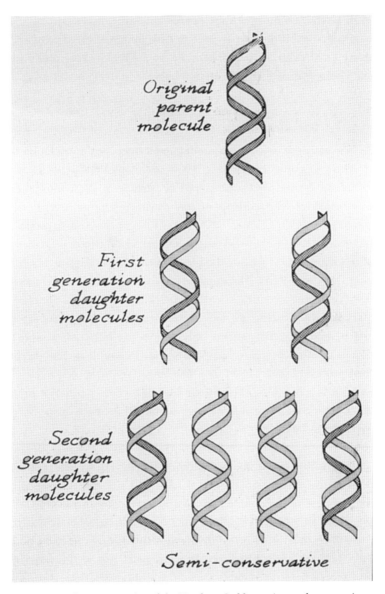

Original parent molecule

First generation daughter molecules

Second generation daughter molecules

Semi-conservative

FIGURE 1.28 The interpretation of the Meselson–Stahl experiment, demonstrating semi-conservative replication (reproduced from L. Stryer, *Biochemistry*, 4th edn, New York: Freeman, 1995).

so the Nobel Committee were spared the difficulty caused by their statutes requiring the limiting of the prize to a maximum of three people. The citation reads 'for the discoveries concerning the molecular structure of nucleic acids and its significance for information transfer in living material'.

The aftermath

As is usually the case with a fundamental discovery, the solution of the DNA structure was only a beginning. Many more questions arose. How did the information carried by the sequence of bases in a DNA molecule finally get transferred into the sequence of amino acids in a protein? The central dogma was formulated by Crick as 'DNA makes RNA makes protein', and this required the genetic code to be worked out (Figure 1.29). It took a further generation of biochemists to work out the biochemical mechanisms involved in transcribing DNA into RNA. The enzyme responsible is RNA polymerase of which there are three varieties in eukaryotes. The enzyme is a complex 'molecular machine' whose structure has been solved by Roger Kornberg (Plates IIIa and b), and this enzyme acts only after a pre-initiation complex, involving dozens of other proteins, has been set up to recruit it. The product RNA is then processed and passed as a messenger from the cell nucleus to the cytoplasm to ribosomes, the protein factories which synthesise proteins of defined sequences. Here the message contained in the sequence of the nucleic acid is translated into a sequence of amino acids according to the genetic code.

I end with a word about the replication of DNA. The principle of semi-conservative replication suggested itself to Crick and Watson directly from the structure of the double helix and is startlingly simple. How, though, does the helix actually unwind, and how does the sequence of each strand get copied? Nucleic acids are only synthesised in one direction (5' to 3'): how then does the antiparallel strand get copied? Again, the work of a generation of biochemists, notably Arthur Kornberg (Kornberg *père*) has shown that it takes dozens of protein complexes, each involving many proteins, to accomplish this. They can be thought of as complex components of several giant molecular machines (Plates IVa and b), which synthesise the new DNA, check it for errors, and pass it on for further interactions which package it in chromosomes.

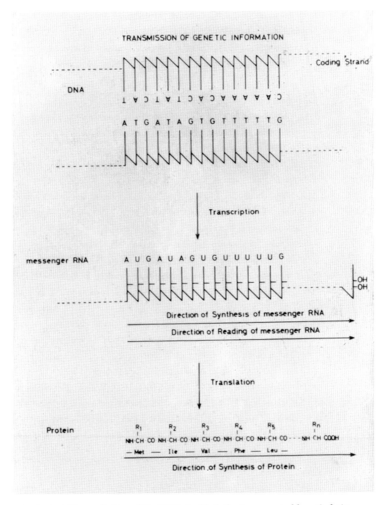

FIGURE 1.29 Transcription of the DNA double helix (represented by a 2-chain ladder) to make messenger RNA, the sequence of which is then translated by the protein synthesis machinery to make a sequence of amino acids, by following the genetic code.

FURTHER READING

Original papers

J. D. Watson and F. H. C. Crick, 'A structure for deoxyribose nucleic acid', *Nature* **171**, 25 April 1953, 737.

M. F. H. Wilkins, A. R. Stokes and H. R. Wilson, 'Molecular structure of deoxypentose nucleic acids', *Nature* **171**, 25 April 1953, 739.

R. E. Franklin and R. G. Gosling, 'Molecular configuration in sodium thymonucleate', *Nature* **171**, 25 April 1953, 742.

J. D. Watson and F. H. C. Crick, 'Genetical implications of the structure of deoxyribonucleic acid', *Nature* **171**, 30 May 1953, 964.

Other relevant papers in chronological order

J. D. Watson and F. H. C. Crick, 'The structure of DNA', *Cold Spring Harbor Symposia on Quantitative Biology* **18**, 1953, 123–31.

F. H. C. Crick and J. D. Watson, 'The complementary structure of deoxyribonucleic acid', *Proceedings of the Royal Society* A **223**, 1954, 80–96.

R. E. Franklin and R. G. Gosling, 'The structure of sodium thymonucleate fibres. I. The influence of water content', *Acta Crystallographica* **6**, 1953 (submitted 6 March), 673.

'Evidence for 2-chain helix in crystalline structure of sodium deoxy-ribonucleate', *Nature* **172**, 25 July 1953, 156.

M. H. F. Wilkins, W. E. Seeds, A. R. Stokes and H. R. Wilson, 'Helical structure of crystalline deoxypentose nucleic acid', *Nature* **172**, 24 October 1953, 759.

R. G. Gosling, thesis, University of London, 1954.

R. Langridge, H. R. Wilson, C. W. Hooper, M. H. F. Wilkins and L. D. Hamilton, 'The molecular configuration of deoxyribonucleic acid: X-ray diffraction analysis', *Journal of Molecular Biology* **2**, 1960, 19.

R. Langridge, D. A. Marvin, W. E. Seeds, H. R. Wilson, C. W. Hooper, M. H. F. Wilkins and L. D. Hamilton, 'The molecular configuration of deoxyribonucleic acid: molecular models and their fourier transforms', *Journal of Molecular Biology* **2**, 1960, 38–64.

Books

J. D. Watson, *The Double Helix*, New York: Athenaeum Press, 1968. Reprinted in G. S. Stent, ed., *The Double Helix: Text, Commentary, Reviews, Original Papers*, New York and London: W. W. Norton, 1980.

R. C. Olby, *The Path to the Double Helix*, Washington, DC: University of Washington Press, 1974; 2nd edn, New York: Dover Publications, 1994.

H. F. Judson, *The Eighth Day of Creation: The Makers of the Revolution in Biology*, expanded edn, Cold Spring Harbor Press, 1996.

S. Chomet (ed.), *Genesis of a Discovery: DNA Structure*, London: Newman Hemisphere Press, 1993 (Accounts of the work at King's College, London.)

A. Sayre, *Rosalind Franklin and DNA*, New York and London: W. W. Norton, 1975.

Other historical references

Wilkins, M. H. F. 'The molecular configuration of nucleic acids', Nobel Lecture in *Les Prix Nobel*, Stockholm, The Noble Foundation, 1962.

A. Klug, *Nature* **219**, 1968, 808–10, 843–4.

M. F. Perutz, M. H. F. Wilkins and J. D. Watson, Reproduction of MRC Report December 1952 and discussion, *Science* **164**, 1969, 1537–9.

F. H. C. Crick, *Nature* **248**, 1974, 766–9.

A. Klug, *Nature* **248**, 1974, 787–8.

H. R. Wilson, *Trends in Biochemical Sciences* **13**, 1988, 275–8, and **26**, 2001, 334–7.

2 Genetic fingerprinting

ALEC J. JEFFREYS

Department of Genetics, University of Leicester

Genetic fingerprinting was accidentally developed in 1984. Initially seen as an academic curiosity, it speedily moved into real-life casework, establishing that molecular genetics really could provide an entirely new dimension to biological identification. This technology has directly impacted on the lives of countless thousands of people involved in criminal investigations, paternity disputes, immigration challenges and the like. This article gives a very personal account of the origins of genetic fingerprinting, of how the technology has evolved over the years into the high-throughput systems now used by law enforcement agencies world-wide, and of how the variable pieces of DNA that underpin genetic fingerprinting can be used to study fundamental aspects of human genetic variability and heritable mutation.

The birth of human genetics

The notion that human characteristics can be inherited, as shown, for example, by family resemblances, has been with us for millennia. However, the study of human genetics in its modern sense can be traced back only to the beginning of the twentieth century, and to just two key individuals. The first was the Austrian physician Karl Landsteiner who in 1900 discovered the ABO blood-group system, the first example of a variable human characteristic that was later shown to be inherited according to the simple rules of inheritance discovered thirty-five years earlier by Gregor Mendel in his experiments on plant hybridisation.

I would like to thank all my colleagues in the Department of Genetics at the University of Leicester, together with many other friends within and outside the University, who have contributed so much to our research efforts. My thanks also to the Medical Research Council and Wellcome Trust who have supported my work over the years, and to the Royal Society which has enabled me to remain a hands-on experimental scientist.

The second was the English physician Sir Archibald Garrod whose work at St Bartholomew's Hospital on the inherited disorder Alkaptonuria provided, in 1902, the first example of a human inherited disease which similarly showed patterns of Mendelian inheritance. These discoveries really defined the subsequent two great research themes in human genetics that continue to this day. The first is the study of normal genetic variation in humans, initially through blood-group variation and subsequently in other macromolecules, using biochemical techniques. These studies provide fundamental information on the genetic variability of the human species and on the diversification of mankind. The second great theme has been understanding how heritable defects in our chromosomes can cause inherited disease.

The birth of forensic genetics

The idea that normal genetic variation in humans could be used to match forensic evidence to suspects is not new. In 1902, Max Richter and Karl Landsteiner suggested that blood-stain typing might be of use in criminal investigations. The development of numerous additional blood-group systems over the following decades, together with the advent of biochemical genetics in the late 1960s which allowed studies of human molecular variation to be extended to a much wider variety of other blood proteins, turned this prophetic suggestion into a reality. Armed with this battery of polymorphic (naturally variable) genetic markers, the forensic scientist could tackle not only blood samples but also other biological specimens, in particular saliva and semen – the latter of course can provide crucial evidence in the identification of sex assailants. These serological and protein markers are also inherited in a simple Mendelian fashion, and can therefore be used to resolve issues of kinship. This branch of forensic genetics reached its zenith in the 1980s with mass application in criminal casework and paternity disputes, and was widely seen by its practitioners as the ultimate in biological identification.

The imperfect world of proteins

While these classic typing systems were in general simple, inexpensive and informative, they had their limitations. Most of the markers show very limited variability – for example, the ABO system can be used to classify people into just four types (blood groups A, B, AB and O), none of which is particularly scarce in the UK. Thus while a mismatch between a suspect and a crime scene sample provides important evidence for exclusion, getting a match hardly constitutes positive proof of a link. The only exceptions are the human leukocyte antigens (HLAs), a complex set of highly variable proteins that lie at the heart of the immune system. HLA markers can be highly discriminating but were at the time rather difficult to type, limiting their usefulness primarily to paternity disputes.

Also, these serological and protein markers were based on complex molecules that all too often became degraded in old forensic samples, rendering them untypable. Further, many were expressed only in blood and could not be used to investigate the many other types of biological evidence all too frequently encountered by the forensic investigator.

With the wisdom of hindsight (a very dangerous sort of wisdom especially in science!), it was obvious that the solution to all these limitations lay in looking at heritable variation, not in the products of genes specifying blood groups and proteins, but directly in the genetic material itself. It was clear, even before the DNA revolution, that the 3 000 000 000 base pairs that make up the human genome must contain many, many sites of heritable variation, offering the possibility of truly positive biological identification. Also, while DNA is a complex molecule, it is surprisingly tough and bits can survive in typable form for remarkably long periods, as discussed by Svante Pääbo in this volume. While the Jurassic Park scenario of truly ancient DNA has now been largely discredited (DNA is tough but not *that* tough!), it is often possible to recover DNA fragments from remains even thousands of years old, providing a fascinating 'time machine' with which to explore the genetic landscape of the past. Finally, human DNA is universal – essentially anything from a human body, whether blood, tissue, semen, hair, nail clippings, saliva, urine or faeces, contains DNA, and it is now clear that even handling an object can leave behind traces of DNA which can be detected using the most sensitive current typing systems.

One small problem though – how could genetic variation in specific DNA sequences be detected amongst the impenetrable tangle of DNA that makes up our genome? The answer lay in the development of an entirely new branch of science.

The rise of genomics

The announcement by Watson and Crick in 1953 of the double helical structure of DNA revealed for the first time the digital nature of heritable information and signalled the beginning of a molecular revolution that is still with us. Recombinant DNA technology, first developed in 1973, allowed complex genomes to be broken down into small pieces that could be propagated in bacteria, allowing individual genes to be studied in detail. The first human gene was isolated in 1977 and today we are close to seeing the complete sequence of the 3 000 000 000 base pairs that constitute the human book of life. This new science of genomics exploring gene and chromosome organisation, while hugely informative about the diversity of information stored in our DNA, nevertheless tells us nothing about genetics as defined as the study of inherited variation – indeed, the human genome project would have proceeded just as well even if we were all clones of each other, with no inherited variation whatsoever! However, genomic technologies have proved crucial in exploring variation between individuals in our precise DNA sequence and it is this theme that has fascinated me over the last twenty-eight years.

Inherited variation in human DNA

Back in 1977, the only tools we had for exploring human DNA variability were restriction enzymes that allowed us to cut DNA at predetermined sites, and DNA probes (radioactively tagged single strands of the double helix) which enabled us to detect specific regions of DNA in human chromosomes by nucleic acid hybridisation (two complementary strands uniting very specifically to reform a double helix). By combining these two approaches, we showed that humans do indeed show variation from person to person in the way that restriction enzymes cut their DNA and that these variations in the genetic material were inherited in a simple Mendelian fashion.

These so-called restriction fragment length polymorphisms allowed us to estimate for the first time just how variable human DNA is from person to person – the answer, at the molecular level, was not much, with perhaps one base pair in every 300 showing polymorphic variation between people. However, at the level of the whole genome, the variation was immense with some 10 000 000 different positions in our chromosomes showing variation. The problem of limited numbers of genetic markers was therefore, in principle at least, solved. The drawback was that these single nucleotide polymorphisms (SNPs) were at the time rather difficult to detect and assay. Also, they are individually not very informative – if a given base shows polymorphic variation with alternative forms (alleles), say A and T, and given that we all inherit two copies, one from each parent, then a given SNP can only be used to classify people into just three types (A/A, A/T and T/T). This is no better than the worst of the old blood-group systems!

Super markers

We therefore started in the early 1980s to search for far more variable regions in human DNA – surely there must exist bits of DNA that show not just two alleles, but instead tens, hundreds, maybe thousands of different versions. The driver behind this research was the need to develop much more informative genetic markers for basic and medical research, in particular for mapping human chromosomes; forensic DNA analysis was certainly not in our thoughts! There were a few reports in the scientific literature that such hypervariable DNA did exist, and consisted of what we subsequently termed minisatellites – regions of DNA consisting of 30 or so base pairs repeated over and over again for tens or hundreds of times, and with different alleles varying in the number of stutters. The problem was how to access these minisatellites.

The clue came from a totally different project that started with a sample of seal meat obtained from the British Antarctic Survey Headquarters in Madingley, Cambridge – we needed this specimen for our work on DNA evolution, in particular of the gene that codes for myoglobin, the oxygen-carrying protein present in huge amounts in seal muscle. Having isolated the seal myoglobin gene, we felt obliged to compare it with its human counterpart (this work was funded in part by the Medical Research Council, so a human dimension seemed sensible). It proved a wise decision. Tucked away inside the human gene was a minisatellite – nothing very exciting until we noticed that the stuttered sequence looked rather familiar, not unlike the stutters in the few other minisatellites described in the literature. Maybe DNA preferred to stutter at a specific sequence, in much the same way that a person will tend to stutter over a specific sound. The implications were clear – if we used a hybridisation probe consisting of this DNA sequence motif shared by different minisatellites, then it should latch onto many different minisatellites simultaneously, giving us unlimited access to these potentially extremely informative genetic markers.

Stumbling upon DNA fingerprinting

The key experiment in September 1984 was a small test of a range of samples that included, for no particularly good reason, DNA from a human father/mother/child trio as well as DNA from a baboon, a seal, a cow, a mouse and a tobacco plant. While the results were hideously murky (Figure 2.1a), it was clear that we were detecting lots of what appeared to be highly variable DNA fragments. Mum and dad were obviously different, and the child seemed to be a union of the parents' DNA patterns. Equally exciting, it looked as if the system worked on other species too.

Within months, we had improved the technology to the point where we could resolve large numbers of extremely variable DNA fragments containing these minisatellites (Figure 2.1b), not just in humans but in other organisms as well. In humans, these banding patterns were clearly highly individual-specific, with the chance of matching even between close relatives or members of an isolated inbred community being essentially zero. Within a given human, though, the patterns were constant, irrespective of the source of DNA, whether from blood, semen, hair roots or whatever. Testing families showed that the multiple markers that make up a DNA fingerprint were inherited in a simple Mendelian fashion, with each child receiving a random selection of about half of the father's bands and half of the mother's. It was clear that we had accidentally stumbled on a DNA method not only for individualisation but also for establishing family relationships. Thus was DNA fingerprinting born.

The first case

While we could immediately see the potential of DNA fingerprinting, we felt that it would take years before it would move into real casework. The one sensible decision we did take was to use the term 'DNA fingerprint' rather than the more accurate, if rather incomprehensible, term 'idiosyncratic Southern-blot minisatellite hybridisation profile'. This decision helped bring our work to public attention through the press, and led to our first enquiry in spring 1985 from Sheona York of the Hammersmith and Fulham Community Law Centre who was representing a family of UK citizens originally from Ghana who were embroiled in an immigration dispute. The youngest son had visited Ghana, then returned to London on a tampered passport; the immigration authorities took the reasonable view that the returning boy was a substitute, either unrelated to the family or the son of one of the mother's sisters in Ghana, and did not grant him residence. The family then went through a full range of classic blood typing, including nine blood groups, seven other blood markers and the relatively informative HLA system, but with equivocal results – it was

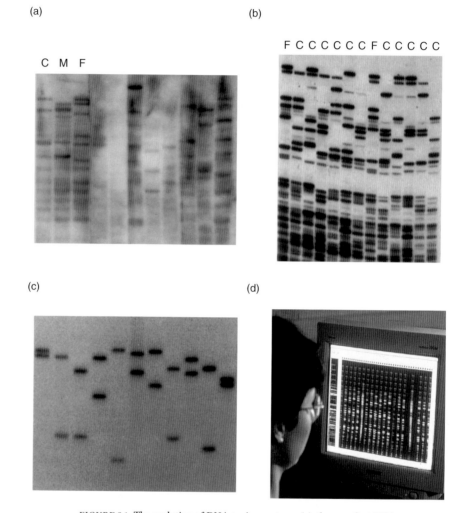

FIGURE 2.1 The evolution of DNA typing systems: (a), the very first DNA fingerprints with a family group at left (M, mother; F, father; C, child) plus DNA from various non-human species; (b), improved DNA fingerprints, from a single family with the father (analysed twice) and his 11 children. Note how DNA fingerprints readily distinguish even close relatives, and how bands in the missing mother can be easily identified as bands in the children that are not present in the father; (c), simpler DNA profiles of unrelated people; (d), DNA profiling using PCR-amplified microsatellites. Several microsatellites are amplified at the same time and the resulting profiles are displayed on a computer and automatically interpreted for databasing (courtesy of Cellmark Diagnostics).

99% certain that the woman and boy were related in some way, but it was unclear whether the woman was the boy's mother or aunt. (As an aside, the written evidence on blood groups actually included a clear exclusion, with the woman being typed as Hp2/2 and the boy Hp1/1, but I assume that this must have been a typographical error.)

Could DNA fingerprinting help this family? The case was difficult – not only were the father and the mother's sisters unavailable for testing but the mother was unsure about the paternity of the boy in dispute. All we had were blood samples from the mother, her three undisputed children and the boy. However, the evidence from the DNA fingerprints was clear – every variable band in the boy could be matched either to one of the mother's bands or to a paternal band present in one or more of the undisputed children (these children allowed the missing father's DNA fingerprint to be reconstructed) (Figure 2.2). The conclusion was that this boy was a full member of the family, and further had the same father as the other children. The statistics were impressive: it was 99.999 999 8% certain that the woman and boy were related, and 99.997% certain that she was the boy's mother rather than his aunt – not proof positive (a mathematical impossibility) but as close as makes no difference. The evidence was submitted to the Home Office – despite the fact that they were only vaguely aware of DNA and certainly had never heard of DNA fingerprinting, they positively reviewed the evidence and dropped the case against the boy, granting him residence in the UK and allowing him to remain with his family. Interestingly, they decided not to cross-examine the DNA evidence in an immigration tribunal; while not a court-of-law, such cross-examination would have established a degree of precedence for the admissibility of this arcane new scientific evidence.

Opening the flood gates

The lesson from this case, namely that here was a new technology that could resolve immigration disputes, did not go unheeded. Within months, we were bombarded with requests for testing, primarily from families from the Indian sub-continent where the husband was resident in the UK and was seeking to be reunited with his wife and children, but did not have documentary evidence such as birth and marriage certificates needed to support his claim. Traditionally, these disputes were resolved by Entry Clearance Officers who would interview the family, relatives and neighbours in an attempt to establish the

Shepherds Bush Gazette 17 May 1985

That's my boy

Mother turns to science in fight for son

A MOTHER is praying that a scientific break-through will mean she can stay in Britain with her son.

Mum Christiana Sarbah hopes the breakthrough will finally convince immigration officials she is the mother of Andrew, 15.

A scientist using the new technique has said the chance of Andrew not being her son is at the most one in 30 million.

Using the "genetic fingerprint" in blood cells from mother and son, the scientist at Leicester University has given hope to Christiana after a two-year fight to prove Andrew is her son.

The fight began the minute Andrew, who was born in England, arrived back in this country after a long stay in Ghana. Immigration officials held him at Heathrow, saying his passport had been forged.

He was allowed to stay at the family home in Fulham Palace Road with his three brothers and sisters after intervention by MP Martin Stevens.

Ever since, Christiana has been fighting the Home Office with the help of lawyers at Hammersmith Law Centre.

The law centre heard of the scientist's work and contacted him. In one of the first practical uses of the

By John Millard

genetic fingerprinting technique, he produced his startling evidence.

Now the law centre is to appeal to the Home Office armed with the findings.

This week Christiana, a nurse, said: "It's terrific news. It's wonderful what technology can do. I'm lost for words.

"People who know me and know Andrew know he is my son. I have gone through hell.

"I don't know what else to give for the immigration authorities to believe me."

She said both Andrew and herself had had their health affected by the struggle. "I used to be a jolly person and enjoy life, but now I'm miserable," she said.

Christiana says she will go wherever Andrew goes.

She praised workers at the law centre, which is threatened with the loss of its grant later this year. "It's cost so much money, this case, but I can't pay them," she said. "But I've started doing the pools."

Christiana Sarbah and son Andrew.

FIGURE 2.2 The first immigration dispute resolved by DNA fingerprinting. The boy was successfully reunited with his mother (cutting courtesy of the *Shepherd's Bush Gazette.*)

validity of the claim. There followed a pilot study with the Home Office on volunteer families, and subsequent very extensive casework by Cellmark Diagnostics, a company established in 1987 to provide a commercial outlet for DNA fingerprinting. Many thousands of families have now been reunited by DNA testing. It soon became clear that the substantial majority of claims were authentic, indicating that significant injustice had been done to many families previously refused admission to the UK. However, DNA testing can result in the occasional very unpleasant surprise. When non-paternity is detected in a family where the man genuinely believes himself to be the father, then the family is placed in double jeopardy, with the child failing to provide the genetic link between husband and wife – leading to refusal of entry – and the family potentially destroyed through the revelation of infidelity. Such cases, though mercifully uncommon, highlight the importance of counselling prior to testing and the difficulty of using a strictly biological rather than social definition of kinship.

The year 1985 also saw this technology take on the first of what proved to be many thousands of paternity disputes. The case involved a divorced English lady, her baby and her 17-year-old French au pair boy. DNA testing was un-equivocal – the boy was the father, with better than 99.999 99% certainty. The evidence was admitted in an English Magistrates' Court, establishing for the first time legal precedence for DNA tests. However, the boy had returned to France, and the subsequent passage of DNA evidence through the French court system was exceedingly protracted and accompanied by complex arguments aimed at establishing the innocence of the boy (including the novel suggestion that he had been framed and that I was the true father – an interesting concept, given that I had never met the woman!). His paternity, however, was eventually accepted.

...And not just people

The fact that DNA fingerprinting could be used on non-human species (Figure 2.3) opened up some very interesting lines of enquiry. Some were obvious – for instance, paternity disputes with dog breeders. Some were less obvious but biologically far more significant. For example, my colleague Terry Burke used DNA fingerprinting to show that sparrows were not quite as faithful as we had assumed – instead, they seemed to indulge in spouse-swapping at a rate quite reminiscent of our own species. Then Esther Signer in my laboratory was approached by Zurich Zoo who held one of the largest captive colonies of the Waldrapp Ibis, a highly endangered species, and were concerned about inbreeding. Using DNA fingerprinting, she managed to reconstruct the entire pedigree of the zoo colony, revealing instances of incestuous (brother/sister) mating that would clearly impact on inbreeding levels. Using this informa-tion, the zoo restructured the colony to minimise this risk of inbreeding, thus helping to maintain levels of genetic diversity. The subsequent impact of DNA fingerprinting on fighting wildlife crime, on understanding breeding systems in the wild and on conservation biology has been substantial.

We have even used this technology to investigate an allegation of scientific error/fraud. The case was that of Dolly the sheep, and the allegation was that

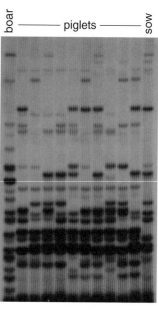

FIGURE 2.3 DNA fingerprints of a pig family. The 'father' is wrong – the piglets contain several paternal bands, absent from the sow, that are not present in the boar, and likewise there are several bands in the boar that are not transmitted to any of the piglets.

there had been a sample mix-up (or, worse, a deliberate fraud) and that Dolly was not a clone. The original publication on Dolly by Ian Wilmut and colleagues did include some DNA typing information to show that Dolly was genetically identical to the ewe from which she had been cloned, though the evidence was limited. However, DNA fingerprinting plus additional more modern DNA markers established beyond any reasonable doubt the genetic identity of Dolly and her donor, proving that she really was a clone. In this context, it is worth stressing that human identical twins also share exactly the same DNA fingerprints – they are after all nature's way of cloning people!

Profiling DNA

DNA fingerprints are excellent for some applications but not for forensic investigations. The patterns are complex and their interpretation is readily open to challenge in court. They are not easy to computer database, preventing direct comparison of samples tested in different laboratories or by the same laboratory on different days. Also, they require significant amounts of good quality DNA, equivalent to that obtained from a drop of fresh blood – many crime scene samples are consequently untypable.

The solution to these problems was simple. DNA fingerprinting not only detects many highly variable minisatellites in human DNA but also allowed us to isolate individual minisatellites using standard recombinant DNA techniques. The original purpose was to use these cloned minisatellites to understand the biological properties of these unusual pieces of DNA. However, the forensic implications immediately became apparent. Each cloned minisatellite, used as a hybridisation probe, produced a much simpler pattern of just two bands per person, corresponding to the two alleles in an individual (Figure 2.1c). These simple profiles could be obtained using considerably less DNA (one hair root is enough), and the estimated lengths of the DNA fragments could be easily databased. DNA profiles revealed the true variability of human minisatellites, with some showing 100 or more different length alleles in human populations. However, DNA profiles are not individual-specific – this is particularly true for siblings, who have a 1 in 4 chance of sharing exactly the same profile, no matter how variable the minisatellite is between unrelated people. Nevertheless, by typing DNA sequentially with a battery of typically five different minisatellites, excellent levels of individual specificity could be obtained – match frequencies of 1 in 1 000 000 000 are entirely typical with DNA profiling.

DNA profiling in a murder case

DNA profiling saw its forensic debut in 1986 with the Enderby case involving two Leicestershire schoolgirls who had been raped and murdered (Figure 2.4). We showed that DNA from the prime suspect – a young man who had confessed to the second murder – did not match semen DNA recovered from both victims. Additional testing by the Home Office Forensic Science Service laid to rest the police's understandable scepticism and resulted in the suspect being freed, the first man to be exonerated by DNA typing. Armed with the DNA profiles of the assailant – the DNA on both victims was from the same man – the police then launched the world's first DNA-based manhunt in the local community. This led to the successful arrest of the true assailant who is now serving a life sentence for both murders.

DNA profiling speedily became the DNA typing system of choice in forensic laboratories world-wide, and remained so well into the 1990s. Typing systems

FIGURE 2.4 DNA profiling in the Enderby murder case (cutting courtesy of the *Leicester Mercury*).

were standardised across Europe and also in North America – though sadly the European and American procedures were different – allowing police to compare databased profiles even between countries. The remarkable discriminating power of this DNA evidence, combined with the simple pictorial nature of DNA profiles that could be readily appreciated and interpreted by juries, ensured countless criminal convictions. However, this huge power had a downside – there was a serious risk that expert witnesses testifying to DNA evidence could subvert (albeit unintentionally) their role as a provider of evidence, and instead give what the jury would interpret as a definitive statement of guilt or innocence.

The challenge

Perhaps not surprisingly, it was US lawyers who led the attack on DNA evidence. The defining case was the Castro murder trial, in which a commercial company provided testimony claiming a clear match between DNA from a murder victim and blood found on the watch of the suspect. However, the defence discovered many serious shortcomings including poor-quality and over-interpreted DNA profiles, lack of appropriate quality control systems, and inconsistencies in the biostatistical evaluation of the evidence. As a result, the evidence was dismissed, though the defendant subsequently pleaded guilty to the murder.

There ensued several years of fierce debate over every aspect of DNA profiling, including the need to establish quality control and assurance procedures and to develop systems that minimise the risk of sample mix-up (though mix-ups will tend to generate mismatches, exonerating the guilty rather than incriminating the innocent). Objective methods were developed for declaring matches between DNA profiles, and extensive population surveys were conducted to see how DNA profiles varied within and between ethnic groups – information essential to establishing the rarity or otherwise of a set of DNA profiles and thus the strength of evidence in a given criminal case. It was also clear that the defence needed to take a more imaginative approach to the interpretation of a DNA match, which after all only establishes a biological link between two specimens, and says nothing about guilt or innocence – these are issues for the court, not science. For example, proof that semen on a rape victim's clothing matches the DNA profile of a suspect does not *by itself* prove that the suspect was the rapist.

The upshot of this debate, which culminated in two major reports from the National Research Council in the USA, was that DNA profiling emerged as a secure, robust typing system that became widely accepted in courts world-wide. Indeed, this new culture of rigour and excellence soon infected other branches of forensic science, leading to improvements in methodology, interpretation and thus the reliability of forensic evidence.

DNA amplification

The major drawback of DNA profiling is that forensic evidence all too often contains too little DNA, or DNA that is too degraded, for typing. This limitation was solved by the development of the polymerase chain reaction (PCR) invented by Kary Mullis, Henry Erlich and others. This exquisitely elegant and sensitive technology enables minute amounts of DNA to be copied over and over again in the test tube, allowing trace biological evidence found at the scene-of-crime to be amplified and typed. It also enables new typing systems to be developed. One, using variation in single bases in DNA (SNPs), soon became widely used particularly in the USA – while fast, its powers of individualisation were limited. A second and much more discriminating system, which has now become standard in forensic and paternity-testing laboratories, is based on very short stuttered regions of DNA called microsatellites that, as with minisatellites, show variation between people in the number of stutters and therefore length of the microsatellite (Figure 2.1d).

These PCR systems analyse very short fragments of DNA, of the order of 100 bases long, and can be used to type even very degraded DNA. The extreme sensitivity of PCR allows typing to be extended to a much wider variety of evidence including nail-clippings, urine, nasal mucus, traces of saliva on cigarette butts and envelopes, and even ancient skeletal remains. Indeed the most recent systems can type DNA at the level of a single human cell (equivalent to analysing 0.000 000 000 006 gm DNA!), and allow analysis of DNA traces transferred by handling, for example on the steering wheel of a stolen car. While such sensitive systems allow volume crime (burglary, car theft) to be investigated, they do pose significant problems in terms of contamination, both at the scene-of-crime and in the testing laboratory, and ought to raise questions about the relevance of such evidence to a criminal case – after all, we all leave behind us a trail of DNA, and we are not all criminals!

Some prominent cases

We first used microsatellites in 1989 in an investigation of skeletal remains exhumed from a grave site in Brazil and believed to be those of Dr Josef Mengele, the notorious Auschwitz concentration camp doctor. By comparing bone DNA profiles with profiles of living relatives of Mengele, it was possible to establish, with 99.9% confidence, that the remains were indeed those of the war criminal. This, together with evidence based on matching the skull to photographs of Mengele, allowed a major war crime investigation that had proceeded for decades to be brought to a successful conclusion.

Another, and truly remarkable, investigation was that conducted by Peter Gill of the Home Office Forensic Science Service and Pavel Ivanov of St Petersburg on skeletal remains recovered from a grave near Ekaterinberg and suspected of being those of Tsar Nicholas, the Tsarina and three of their children murdered by the Bolsheviks in 1917. Microsatellite typing confirmed that they did indeed constitute a family group, but which family? The solution lay, not in microsatellites that can only be used to determine immediate kinship, but instead in mitochondrial DNA. This small DNA is found in mitochondria, the power-houses of cells, and is unusual in being inherited strictly from mothers to children, with men making no contribution. Matrilineal descendants of the Tsar and Tsarina were traced, including the Duke of Edinburgh, and typed to show that the mitochondrial types at the grave site did indeed match those predicted for the royal family. Statistics raised its head, however – could this be some other Russian noble family with an overlapping maternal ancestry? With no database of Russian nobility mitochondrial types, the question remained unresolved. However, the Tsar's DNA sequence contained a clue – one position showed a mixture of two different bases. This unusual condition, termed heteroplasmy, is due to a new mutation shared by some but not all of the thousands of mitochondrial DNA molecules present in each of our cells. Since this mixture usually takes a few generations to become resolved, it was possible that the same mixed mutation might exist in other close matrilineal relatives of the Tsar. So the Tsar's brother, Grand Duke Georgij Romanov, was exhumed and typed, revealing exactly the same heteroplasmy. The case was clinched, allowing the Romanov remains the dignity of a state funeral 81 years after the family's murder.

The O. J. Simpson trial in 1995 brought DNA into everyone's living room. Despite a veritable mountain of DNA profiling and PCR evidence, he was acquitted (though a later civil action against him proved successful). This trial provided object lessons in how to attack DNA evidence, particularly from the perspective of possible sample mix-up and the planting of evidence, and exposed the deep gulf that exists between the cultures of science – which works by synthesis and integration of evidence – and of the adversarial legal system which operates more by dissection and reduction. Whether this gulf can be bridged remains a challenge for both science and law.

The President Clinton – Monica Lewinsky investigation was also hugely successful in bringing DNA into the limelight. Astonishingly, information on Clinton's DNA profile became publicly available on the Internet, leading to several women getting themselves and their children tested to see whether paternal characteristics in the child matched those of the President – a curious modern reworking of the Cinderella fairy-tale with DNA taking the place of the glass slipper. DNA test results in the USA are now public entertainment, with alleged fathers being confronted with paternity evidence live on prime-time television. Sadly, this sorry spectacle can now be seen on UK television too – so much for the British sense of decency and fair play.

Databases

Microsatellite typing has been greatly refined and largely automated over the last decade, allowing in 1995 the establishment in Britain of the first national criminal DNA intelligence database. This database currently holds 2 000 000 DNA profiles from convicted criminals and unsolved casework, and has already proved to be extraordinarily effective in the fight against crime, with over 180 000 cases of a prime suspect being identified by trawling the database with a crime-scene DNA profile, and over 15 000 instances of matches between different crimes, as in, for example, serial rape. Similar databases have been, or are being, developed in many other countries. These databases should, however, be seen as an investigative aid and not a prosecution tool. This concept is adhered to in the UK, though I do note that follow-up DNA analysis on a suspect still uses the same ten microsatellite markers as employed in the database, raising concerns about the possibility of a false match – which will steadily increase as the database becomes ever more huge. Surely a safer procedure would be to use different markers in subsequent investigations to provide a genuinely independent verification of the match.

Even with the current database, many crime-scene samples do not yield a prime suspect. The police are therefore considering two approaches that might give a lead in such investigations. The first uses markers on the Y chromosome – this chromosome is strictly inherited from father to son and carries the testis-determining gene. Since surnames (at least in the UK) are also usually inherited from the father, it follows that correlations should exist between names and Y chromosome types and thus that Y typing could give clues about a male perpetrator's surname. Indeed, this concept of patrilineage testing has been used successfully by my colleague Mark Jobling and others to establish that President Thomas Jefferson's Y chromosome was present in a male descendant of Sally Hemmings, one of his slaves, providing substantial support to historical speculation that the President had fathered a child by Sally. However, current evidence suggests that correlations between names and Y types in the UK are not that strong – non-paternity, adoption and the invention of the same name on several occasions during history all conspire to blur the link between a man's name and his Y chromosome.

The second possibility is to scour crime-scene DNA for markers that might give clues about the physical characteristics of the perpetrator – ethnic group, skin colour, hair colour (red hair is already testable), eye colour (partially predictive tests exist) and perhaps even facial features (though we still have no clue as to how or which genes control our appearance). The problem of course is whether the public would accept such prying into genetic markers that are important to us as individuals, in contrast to the currently used microsatellites which contain no information on appearance and only the weakest of connections with disease risk.

There is every likelihood that Y chromosome and physical characteristic testing will yield only the feeblest of clues in criminal investigations where there is no suspect. However, there is an alternative and much more scientifically robust approach, namely expanding the current DNA database. Already, police in the UK can retain DNA from suspects who are subsequently cleared in an investigation (the logic here seems to be that such suspects are more likely than randomly selected individuals to commit a crime in the future, though I have yet to see any evidence in support of this concept). The obvious extension would be to extend typing to the entire population, perhaps running a global database in parallel with the presently configured criminal database but held by an agency quite separate from the police. In this sense, DNA could be seen as providing a certificate of identity – analogous to certificates of birth, marriage and death – not only serving as a deterrent against crime but also of potential positive benefit to individuals and their family. One clear benefit would be in mass disaster analysis, for example the terrorist outrages in the USA and Bali. The Twin Towers disaster generated some 25 000 body parts requiring identification by DNA, a distressing task made worse by the lack of reference DNA profiles and the resulting need for investigators to search the homes of suspected victims for personal items that might carry their DNA. While a global database merits serious consideration, and has precedence in the large DNA databases being constructed by the military to provide molecular 'dog tags' of their personnel, it is far from clear whether the right structures and safeguards could ever be developed to make it publicly and politically acceptable.

Detecting mutation

The highly variable minisatellites used in DNA fingerprinting have also allowed us to study fundamental issues concerning the nature of human DNA variation, in particular in trying to understand the mechanisms which result in heritable changes in our DNA. There are two key processes at work. The first is mutation whereby spontaneous, or possibly environmentally induced, alterations in our DNA become established and transmitted to subsequent generations; unfortunately, mutation occurs over most of our DNA at a frequency so low as to be undetectable in human families. The second process is recombination or crossing over, in which maternal and paternal chromosomes pair up during a specialised cell division (meiosis) and exchange genetic material before being separated into individual sperm or eggs. Recombination is fundamentally important in reshuffling patterns of variation in human chromosomes and greatly increases the genetic diversity of mankind, exposing novel combinations of genetic markers to the forces of natural selection. Again, our knowledge of processes of recombination in human chromosomes is severely limited by the very low rate, at the molecular level, at which this process occurs. The challenge therefore has been to develop new approaches to the very high-resolution analysis of mutation and recombination in an attempt to study the evolution of human DNA in real time.

Minisatellites have proved to be a superb test-bed for developing these technologies. These regions of DNA are not only extremely variable but can also mutate (change their number of repeats) at a phenomenal rate, with in some cases as many as 1 in 10 sperm or eggs carrying a new mutation, compared with perhaps 1 in 1 000 000 for mutation in conventional genes. Given this instability, it is often easy to detect heritable mutations even in small families – such mutations are an irritation in paternity testing but very useful for our purposes. Even better, it is possible to use PCR amplification of single DNA molecules to detect new mutations directly in sperm DNA, enabling unlimited numbers of germline mutants to be recovered from any appropriate man (Figure 2.5).

Mutation and recombination

By comparing the structure of a minisatellite before and after mutation, we have found, rather surprisingly, that minisatellite mutation is in fact driven by the process of crossing over at meiosis; thus for this class of DNA at least, mutation and recombination are not different and distinct processes. In most cases minisatellites mutate by the copies (alleles) on homologous chromosomes pairing, followed by information being copied from one allele and pasted into the other, resulting in a recombinant mutant but without crossover as such. However, using new and very powerful single-sperm techniques it has been possible

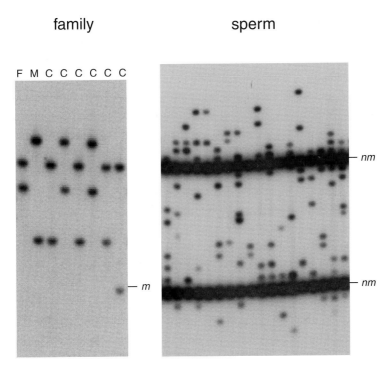

FIGURE 2.5 Detecting minisatellite mutation in families and in sperm. The DNA profiles of the family show bands transmitted from father (F) and mother (M) to the children (C) according to Mendel's rules of inheritance, with each child inheriting one or other of each parent's bands at random. The last child shows a maternal mutation (*m*). The sperm typing involves amplifying DNA from 100 sperm per lane. Most of the sperm carry one or other of the non-mutated alleles (*nm*), though many sperm mutations which alter the length of the minisatellite are visible. For this minisatellite, about 1 sperm in 16 carries a mutation in this man. To detect this number of mutations by family analysis would require this man to have produced 2000 children!

to study true crossovers at very high resolution. This has shown that minisatellites are located at the boundaries of intense crossover hot spots which seem to drive instability. It therefore seems that minisatellites are hot spot parasites that have succeeded in engaging the recombination machinery to propagate themselves. Whether other classes of stuttered DNA, such as microsatellites, evolve by the same process remains unclear – the current strong suspicion is that different processes such as aberrations in DNA replication or repair may be important drivers of microsatellite instability.

Hot spots and DNA diversity

The discovery of crossover hot spots near minisatellites raised major issues about the distribution of crossover events along human chromosomes, long thought to be fairly random. This is not just an academic query – the clustering of crossovers into hot spots could profoundly influence the way in which variation is distributed along human chromosomes, with major implications for the search for genes involved in human inherited disease. We are therefore surveying other regions of the human genome, in particular in the Major Histocompatibility Complex (MHC), a giant complex of genes intimately involved in the function of the immune system. This region has long been studied in families, in particular in relation to tissue matching in transplantation surgery, and these family studies have provided some evidence that crossovers may be significantly non-randomly distributed in this region of the human genome.

By combining studies of patterns of DNA diversity in populations with very high-resolution analysis of crossovers in single sperm, we find that highly localised hot spots do appear to be the rule, at least in the MHC. This implies that crossovers in human chromosomes are highly targeted – though what controls this targeting remains completely unclear. These hot spots can exert a huge influence on patterns of DNA diversity, organising DNA into largely non-recombining segments that behave as if they were in a non-sexual organism, separated by hot spots in which sexual recombination has heavily scrambled patterns of diversity. It is also becoming clear that these hot spots can behave in bizarre ways, breaching the fundamental laws of Mendelian inheritance and providing genome-driven mechanisms that can propel new sequence variations into human populations.

Mutation and radiation

It has long been known that environmental agents such as ionising radiation can induce inherited mutation in the DNA of experimental organisms. Surprisingly, there is no evidence for such a phenomenon in human populations, despite major surveys, particularly of individuals exposed to radiation following the Hiroshima and Nagasaki atom bomb attacks. The problem lies in the very low mutation rates associated with traditional germline monitoring systems, which require surveys of hundreds of thousands, or indeed millions, of families to be able to detect induced inherited mutation. My colleague Yuri Dubrova and I are therefore using highly unstable minisatellites as a novel approach to mutation monitoring, in particular to see whether they respond to environmental agents such as ionising radiation.

Irradiating mice

Experimental exposure of mice to low doses of x-rays, γ-rays or neutrons showed, surprisingly, that these unstable regions of DNA do respond remarkably to radiation, resulting in an easily detectable increase in new mutants appearing in the offspring of exposed animals. Some features of induction in mice are similar to radiation-induced mutation at real genes, detected for example by the appearance of coat colour mutations (only detectable in huge numbers of offspring mice). However, other features strongly suggest that radiation induces minisatellite mutation by a novel and bizarre process which does not involve radiation causing damage at the minisatellites themselves. Instead, it appears that exposure creates an as-yet-unidentified signal in the germ cells of mice, and that this signal is activated at some stage to cause genome-wide destabilisation of repeated DNA. Remarkably, radiation-induced mutants appear not only in the offspring of exposed mice, but also in their offspring and grand-offspring in turn, even though they have never been exposed to radiation. This extraordinary phenomenon of transgenerational mutagenesis, whereby a mouse can 'remember' that its parent or grand-parent was exposed to radiation, indicates that the exposure signal can be stably inherited down the generations. This obviously raises major issues concerning the long-term effects of radiation exposure.

...And humans

Such exposure experiments are of course very difficult to perform in humans. We have therefore focused our attention on populations accidentally exposed to radioactivity. In a major study on the Chernobyl disaster, we have shown that populations in Belarus and the northern Ukraine living in contaminated regions show an unusually high mutation rate in their children. Further, the mutation rate appears to correlate with the levels of radioactive fallout in their environment. We have also seen a similar phenomenon in populations exposed to radioactive fallout in the vicinity of the former Soviet nuclear weapons testing facility at Semipalatinsk in Kazakhstan.

These studies provide the first direct evidence for radiation-induced inherited mutation in humans, but must be treated with caution since there may be other confounding environmental factors responsible for shifting mutation rates. Two studies on the Chernobyl liquidators, the people sent in to clean up the power station following the disaster, have produced conflicting results, one showing some evidence for radiation-induced mutants appearing in their children and the other showing no effect. Minisatellite mutation studies on Japanese atom bomb survivors and their families have shown no evidence for radiation-induced mutation, nor is there any evidence that radiation induces sperm mutation in men undergoing lower abdominal radiotherapy – this latter is a well-controlled experiment, closely analogous to the mouse irradiation work, which involves analysing single-sperm mutation rates on samples provided by patients before, during and after testis irradiation.

Clearly, much more work is needed to see why human minisatellites appear under some circumstances to respond to radiation in the remarkable fashion seen for repeat DNA in mice, yet in other circumstances show no obvious response. One possibility is the nature of exposure – all the positive studies involve populations subjected to chronic low-level exposure to radiation present in the environment, including food and drink, while the negative results are united by the common factor of external, relatively acute exposure. If radiation-induced mutation in humans is verified, then it clearly has major implications both for understanding the interaction between radioactivity and the germline and for radiobiological protection.

The future

With the human genome sequence close to completion, human genetics is already moving rapidly in two major directions. The first is functional genomics, which is attempting to understand the roles of the tens of thousands of genes present in our chromosomes and how they and their gene products cross-talk to create the enormously complex regulatory networks that underpin living systems. The second challenge is to gain a global picture of diversity in the human genome and to use this to understand our evolutionary origins and to begin to dissect the many and varied processes of mutation, recombination and natural selection that drive variability into human populations. In the latter respect, we have made our first few steps with minisatellites and recombination, and the future challenge is to address other, rarer modes of genome change that can drive the appearance of biologically more important types of inherited mutation.

FURTHER READING

Alec J. Jeffreys, Victoria Wilson and Swee Lay Thein, 'Individual-specific 'fingerprints' of human DNA', *Nature* **316**, 1985, 76–9. The first description of DNA fingerprinting.

Alec J. Jeffreys, John F. Y. Brookfield and Robert Semeonoff, 'Positive identification of an immigration test case using DNA fingerprints', *Nature* **317**, 1985, 818–19. The Ghanaian family case.

Lorne T. Kirby, *DNA Fingerprinting: An Introduction*, New York: W. H. Freeman and Co., 1992. For those interested in the technical details of pre-PCR DNA fingerprinting and profiling.

Joseph Wambaugh, *The Blooding*, New York, NY: Bantam Books, 1989. An excellent popular account of the Enderby murder case.

The Innocence Project (www.innocenceproject.org). Founded in 1992 by Barry C. Scheck and Peter J. Neufeld to use DNA to exonerate convicted individuals. Lawyers Scheck and Neufeld played a key role in the Castro case and in O. J. Simpson's defence.

Peter Gill, L. Pavel, P. L. Ivanov *et al.*, 'Identification of the remains of the Romanov family by DNA analysis', *Nature Genetics* **6**, 1994, 130–5. How PCR was used to identify Tsar Nicholas and his family.

Bruce S. Weir, 'DNA statistics in the Simpson matter', *Nature Genetics* **11**, 1995, 365–8. A poignant if technical account from a lead prosecution expert.

Forensic Science Service Online (www.solvethecrime.co.uk/forensic/entry.htm). A useful site with frequent updates on DNA-based criminal investigations and on the progress of the UK National DNA Database.

3 Ancient DNA

SVANTE PÄÄBO

Max Planck Institute for Evolutionary Anthropology, Leipzig

The vast majority of all organisms that have ever existed on our planet are either dead or both dead and extinct. However, a tiny fraction of them still exist as specimens collected by zoologists or as finds excavated by archaeologists and palaeontologists. In this essay I will discuss the analysis of ancient DNA from some such remains.

I will first focus on the damaged condition of ancient DNA and the technical challenges that arise from such damage. I will then review some studies that have been done using ancient DNA. One area in which the field has met with great success is in deciphering the genealogical relationships (phylogeny) of extinct organisms. As an illustration I will discuss the genetic relationships of moas, the marsupial wolf and Neanderthals. A more recently emerging line of inquiry has to do with changes over time in frequencies of gene variants within lineages. An example is the genetic changes associated with domestication. Here, I will use maize as an illustration. Finally, I will discuss some prospects for future research.

Post mortem degradation

Research on ancient DNA goes back to the beginning of the 1980s when technologies became available to clone DNA efficiently in bacteria, i.e. to take pieces

My work is funded by the Max Planck Society, the Bundesministerium für Bildung und Forschung, and the Deutsche Forschungsgemeinschaft. I am extremely grateful to many students and postdocs who have produced most of the data mentioned. In particular, Michael Hofreiter, Viviane Jaenicke, Melanie Kuch, Hendrik Poinar, Nadin Rohland and David Serre are currently working in our lab on the projects described. I am indebted to Ann-Charlotte Runn, Uppsala, for expert help with histology; to Darwin College for their hospitality; to Torsten Krude for the editing of the transcript of my lecture; and to Linda Vigilant, Johaunes Krause and others for additional editing of my ramblings.

of DNA from other organisms and replicate them in bacteria. It then became an interesting question to ask whether DNA might survive in specimens such as well-preserved ancient Egyptian mummies that are a few thousand years old. To approach this, I started out by imitating Egyptian mummification by drying samples of human muscles, and then rehydrating them and looking at them in a microscope. This treatment results in morphologically very well-preserved muscle fibres, containing among other easily recognisable structures cell nuclei where the vast majority of the DNA in the cells is stored (Figure 3.1a). When I rehydrated and examined muscles from ancient Egyptian mummies in the same way, however, they usually looked very different. The muscle fibres were barely visible, and no intracellular structure and no indications of cell nuclei could be seen (Figure 3.1b). Probably no DNA at all survives in such mummies. Fortunately, although the majority of mummies look like this, not all do.

For example, a skin sample taken from the left thigh of a 2700-year-old mummy of a child (Figure 3.2) looked quite well preserved. In the basal layers of the skin, structures that looked like cell nuclei were visible (Figure 3.3), and by using a stain that specifically binds to DNA, I detected the presence of DNA in this specimen. So I went ahead and extracted DNA from this tissue as one would do from any modern tissue. The results suggested that human DNA is present in this specimen.

For comparison, I also prepared a series of DNAs of different ages, starting out with a 3-year-old dried piece of meat that we happened to have in the lab. I continued with a 110-year-old marsupial wolf from a zoological museum, with Egyptian and Peruvian mummies that were a few thousand years old and, finally, a 12 000 year-old ground sloth. I analysed the size of these DNA samples by a technique called gel electrophoresis in which an electrical current is applied to a mixture of DNA to separate the DNA pieces by size. All of the ancient samples were degraded to a small size, around 100–200 base pairs, whereas intact modern DNA would be of a size of many tens of thousands of base pairs. Interestingly, this degradation of ancient DNA does not seem to correlate with how old the remains are. The 3-year-old meat is not much less degraded than the 12 000-year-old ground sloth. Therefore, this reduction in size appears to happen quite rapidly after death, and is a general problem with ancient DNA samples.

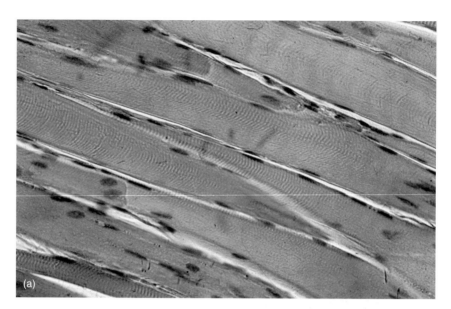

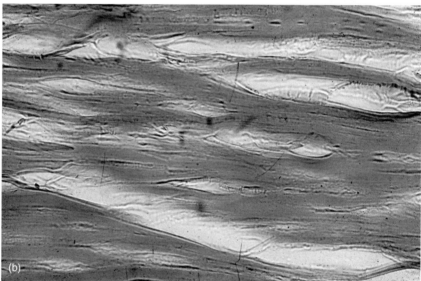

FIGURE 3.1 Microscopic pictures of a human muscle that has been air-dried to imitate mummification in ancient Egypt (**a**) and of a 2000-year-old ancient Egyptian muscle (**b**). Samples were rehydrated and prepared for microscopy in an identical way.

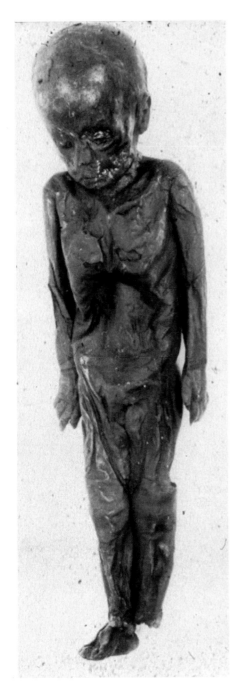

FIGURE 3.2 A 2700-year-old Egyptian
mummy of a child from which DNA was
extracted from the left thigh (photo:
Ägyptisches Museum, Berlin).

FIGURE 3.3 Microscopic preservation of the skin of the mummy shown in Figure 3.2.

DNA damage

The reduction in size or degradation is not the only way in which ancient DNA is modified. After hydrolysis of DNA into its chemical components one can separate and identify the four bases, adenine (A), thymine (T), cytosine (C) and guanine (G) that are the building blocks of the information encoded in the DNA. Modern DNA contains all of these four different components in balanced proportions. In contrast, DNA from the ancient samples contains predominantly G and A, but T and C have largely disappeared. I speculated that the disappearance of C and T is due to oxidative damage, because C and T are particularly sensitive to oxidation. Indeed, when we analysed these DNA samples by mass spectrometry we saw that a large proportion of the natural bases was modified by oxidation. Therefore, we looked more in detail into what types of damage can happen to a DNA strand over time. In addition to oxidative damage, there is a lot of hydrolytic damage that is caused by spontaneous reactions in the presence of water. There is also condensation where cross-links are formed between DNA molecules and proteins and between DNA molecules themselves, and there are other types of damage. Indeed, when we looked at ancient DNA directly in the electron microscope we saw that most of it was degraded to short pieces and we also saw cross-linked structures.

Thus, the major characteristics of ancient DNA include scarcity (it is not present in all old samples, and even when detected is in low quantity), degradation (meaning the DNA is in short pieces) and damage, i.e. the scarce short pieces of ancient DNA that exist are chemically modified (Figure 3.4a). All of these features, as well as contamination with modern DNA, make analysis of ancient DNA particularly challenging.

Generating modern copies of ancient DNA

The use of the polymerase chain reaction (PCR) to make copies of specific DNA segments of interest has revolutionised molecular biology, and made the field of ancient DNA research possible. The PCR process requires short synthetic pieces of DNA that flank the segment of DNA one is interested in. These pieces (or primers) are added to a reaction mixture containing some starter copies of the target (or template) DNA. This mix is then denatured by heating so that the two DNA strands come apart and the synthetic primers are able to bind to their respective complementary sequences. Starting at these primers, two new strands are made on the old DNA as a template by a DNA polymerase. This reaction cycle is then repeated 20–30 times and millions of copies of the piece of DNA one is interested in are created. However, when we perform a PCR reaction on ancient DNA, we encounter four types of problems.

The first problem is that most remains contain very little or no ancient DNA. This means that we can only amplify DNA sequences that occur in many copies per cell and therefore have a higher chance of being present in an ancient sample. Therefore, much of the work that has been done on ancient DNA, and much of the work I will discuss below, involves mitochondrial DNA (mtDNA). This DNA is not located in the cell nuclei but in mitochondria outside the cell nucleus. There are a few hundred or thousand copies of the mitochondrial genome in each cell, whereas each cell contains only two copies of any particular gene in the nucleus.

A second problem is that it is only possible to study short pieces of DNA due to the degradation and damage present. Using the PCR, we cannot generally amplify pieces longer than 200 to 400 base pairs. Consequently, to study a longer DNA sequence that existed in an ancient organism one has to retrieve short overlapping pieces and use these to reconstruct the longer DNA sequence (Figure 3.4b).

S. Pääbo

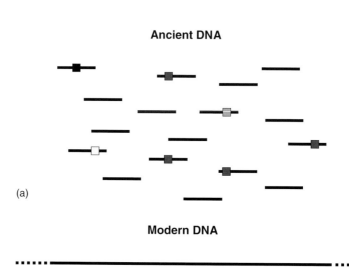

(a)

(b)

FIGURE 3.4 (a) Schematic illustration of the degraded and damaged nature of ancient DNA (below) as compared to modern DNA (above). (b) In order to reconstruct a longer sequence, several short overlapping segments need to be retrieved multiple times by PCR. DNA damage is symbolised by rectangles.

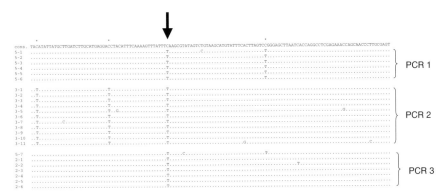

FIGURE 3.5 Individual clones from a late Pleistocene cave bear from which three independent PCRs were performed. As can be seen, while three substitutions are seen in only one PCR each, they are inferred to be due to errors by the DNA polymerase during the first or early cycles of the amplification process. Only one substitution (arrow) is seen in all PCR and is therefore inferred to represent the authentic DNA sequence that this individual carried when alive.

The third type of problem encountered in PCR reactions from ancient DNA is caused by the accumulated damage. For example, when we amplify a piece of ancient DNA, and clone individual molecules for sequence analysis, we often find a particular base substitution in all of the amplified molecules, in comparison to a reference sequence. However, if we repeat this procedure from the very same DNA extract, the previously seen base substitution is no longer seen and other ones have appeared at other sites. When we repeat this procedure again, all previously seen base substitutions may have disappeared (Figure 3.5). The reason for this variability lies in the fact that the amplification by PCR has started from single molecules and that damage in that starting molecule can cause the DNA polymerase to insert the wrong base at the site of damage. Other types of damage can cause additional problems. Consequently, one needs to repeat each experiment several times to determine the true base for each position in the DNA sequence.

Contamination

The fourth and final problem is a grave one that haunts the entire field – the contamination of specimens and experiments with contemporary DNA. It is well illustrated by the following story: we extracted DNA from a skeleton of a moa, an extinct flightless bird that lived in New Zealand. We tried to amplify a

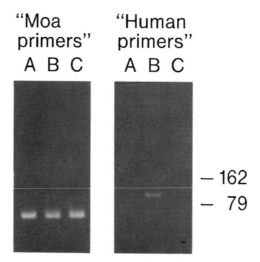

FIGURE 3.6 Amplification from two extracts (Lanes A and B) of a moa bone. Lane C represents a control extract to which no bone was added. Using primers that will amplify a DNA sequence from this group of birds, no amplification products are obtained (the bands seen near the bottom are primer artefacts). When primers that amplify a human DNA sequence are used, one extract yielded a product. The contamination of ancient samples with traces of contemporary human DNA is very common and makes the study of ancient human remains difficult and often impossible.

short piece of this DNA using synthetic primers that we knew would only work for the DNA of moas. We failed to amplify anything. We then used primers designed for human DNA. One of the extracts produced a nice band in a PCR amplification, which we sequenced and found to be a human sequence (Figure 3.6). This does not mean, however, that these birds are particularly closely related to humans. Instead, it indicates that some museum curator or palaeontologist had been handling the specimen and that we had amplified the DNA from that person.

Criteria of authenticity

As a consequence of these problems, we have for the last fifteen years recommended adherence to a set of criteria that in our view are required in order to believe the results from analysis of ancient DNA, including work done in our laboratory and by others. These criteria are straightforward applications of good scientific principles, and include: (1) inclusion of appropriate negative controls; (2) chemical analysis of specimens to make sure it is reasonable that authentic endogenous DNA could survive in them; (3) replication of results, in other laboratories, when the results are of a particularly startling nature. Unfortunately, not all published analyses of ancient DNA are reliable due to violations of one or several of these basic principles.

Museum collections

When one adheres to these criteria, it is actually possible to study thousands of specimens in museums and in other collections. Here, I will briefly mention some such projects, covering an increasingly greater time-scale. The questions that are addressed will range from shifts in population history of mammals over the last hundred years to the phylogeny of extinct animals and its implications for how their morphology has evolved and how they have spread between continents.

To follow the history of animal populations over the past 100 years, one can use field maps from the beginning of the twentieth century to identify the location where zoological specimens were collected. One can then go into the field and put out traps again within meters of where the traps were put, say, 70–80 years ago, and catch modern specimens. This allows a direct genetic comparison between the old and the contemporary populations. One can then determine if the animals living in an area today are the descendants of the individuals that lived there in the beginning of the twentieth century, or if major population movements have taken place. For the first species where this was done, kangaroo rats in the Californian desert (Figure 3.7), the situation proved to have been very stable over the last 50–80 years, since old and new populations were very closely related to each other. Similar studies have now been performed for a number of species such as pocket gophers, field mice and rabbits. In most cases, populations have been stable over time but in some cases dramatic and previously unknown replacements have been documented.

Going back further in time, there have been many molecular studies where the relationships between extinct and living species have been determined. This work began with recently extinct animals such as the Tasmanian wolf in

FIGURE 3.7 Kangaroo rats from California, trapped in 1917 (left) and in 1988 (right), from which mtDNA sequences were retrieved. The results showed stability of the population over the time period investigated (photo: F. X. Villablanca).

Australia. This is a marsupial animal that looked similar to the placental wolf in Eurasia and the Americas. It became extinct around 100 years ago through hunting. When we extracted DNA and compared it to other marsupials, we found that the marsupial wolf is closely related to other Australian marsupial animals, and not closely related to carnivorous marsupials in South America although it looks superficially similar to those. So the marsupial wolf is a classical example of convergent evolution at the morphological level in two regards. It not only looked similar to a placental wolf to which it is not related but had also evolved to look similar to South American carnivorous marsupials to which it was also not closely related.

An early example of an older extinct species whose relationships with extant species were clarified are the moas in New Zealand. These were big flightless birds and became extinct a few thousand years ago. DNA has by now been extracted from many moas and compared to existing flightless birds. The big surprise that came out of early studies was that the kiwis that exist today in New Zealand (the relatively small flightless birds, not the people) are not closely related to the extinct moas. Rather, the moas were more closely related to Australian flightless birds. These experiments were done by Alan Cooper, who has since gone on to be a professor at Oxford. Recently, he has sequenced the complete mitochondrial genome of about 16.5 kb of DNA from several species of moas – the longest ancient DNA sequenced to date. Using these sequences, he has been able to date the divergence between moas and other flightless birds to approximately 82 million years ago whereas the divergence of kiwis from the other flightless birds dates back about 68 milllion years. This indicates that the kiwis came to New Zealand after it had separated from Australia. Therefore, they must have colonised New Zealand over water.

Even older extinct species that we have studied are, for example, giant ground sloths that are up to 50 000 years old. There will, however, be a limit to how far back in time we can go. Based on our knowledge of the chemical processes affecting the DNA, we can estimate that limit to be somewhere within the last million years. The oldest believable DNA sequences determined to date are just beyond 50 000 years of age, for example from mammoths in the permafrost. DNA sequences older than that, for example from dinosaur fossils or from insects in amber, have not been reproduced and thus fail the criteria of authenticity mentioned above. There are indications that we may be able to extract and analyse DNA a bit older than 100 000 years, perhaps even going back half a million years, but the limit is surely on this side of a million years. Within that time span, however, there has recently been an explosion of studies of the kind I have just discussed.

Neanderthal genetics

I will now focus on two studies that we find particularly interesting at the moment. The first concerns the question of the relationship between anatomically modern humans and the Neanderthals. Human forms evolved in Africa and appeared outside of Africa a little after 2 million years ago in the form of *Homo erectus*, and evolved further in Europe into Neanderthals. Classical Neanderthals appeared around 100 000 years ago and existed until around 30 000 years ago. Anatomically modern humans appeared in the fossil record first in Africa and the Middle East around 100 000 years ago and came to Western Europe about 50 000 years later. Since modern humans thus co-existed in Eurasia with the Neanderthals until they disappeared, an interesting question concerns the genetic relationship of modern humans with Neanderthals. Is

FIGURE 3.8 Cranium of the Neanderthal type specimen from which mtDNA sequences were determined in 1997 (photo: R. W. Schmitz).

there genetic continuity between Neanderthals and modern humans, as one view suggests? Or did the modern humans that came out of Africa have no interactions with the Neanderthals until the latter disappeared? Was there a lot of mixing or not?

In 1997 we were happy to be allowed to remove half a gram of bone from the right upper arm of the Neanderthal type specimen that was found in 1856 in Neanderthal in Germany (Figure 3.8) and gave its name to that group of hominids. We extracted DNA and reconstructed a piece of a hyper-variable part of the mitochondrial genome as described above. This allowed us to compare for the first time a Neanderthal DNA sequence to contemporary human DNA sequences. Within present-day living humans from around the world, there was a mean value of around 8 differences (base substitutions) between the sequences. When comparing this to the Neanderthal sequence, the mean was around 27 differences. Furthermore, in tree reconstructions that estimate the evolution of this piece of DNA, the Neanderthal falls clearly outside of the variation of modern humans.

All contemporary humans trace their ancestry in the mitochondrial DNA back to a common ancestor that lived in the order of 100 000–200 000 years ago. The common mitochondrial ancestor that we share with the Neanderthals existed further back, about half a million years ago. Therefore, this analysis tells us unequivocally that Neanderthals did not contribute mitochondrial DNA to present-day humans. However, it does not exclude that some genetic mixture occurred in the Late Pleistocene since a Neanderthal mitochondrial DNA contribution could have became lost by the chance events associated with the fact that some segments are passed on, whereas others become extinct in each generation. However, from the direct study of Neanderthal DNA sequences, there is no positive indication of gene flow between the two groups. Other work on contemporary human genetic variation, for example on the Y chromosome and in other parts of the nuclear genome, similarly yielded no evidence of a contribution of Neanderthals and other archaic humans to contemporary humans.

It should be noted, though, that the genetic data have often been over-interpreted, for example to say that there was no intermingling, no common background, or no connection at all between Neanderthals and modern humans. That is not true. As I have argued above, we cannot exclude that there had been some mixing going on in the late Pleistocene. Furthermore, even without mixing, modern humans were not that different from Neanderthals. We can infer this from the fact that we can date the divergence for *mitochondrial* DNA to somewhere around half a million years ago, while at the same time some genes in the contemporary human *nuclear* genome trace the ancestry of their variation as far back as a million years and more. Therefore, if we were able to study nuclear DNA in Neanderthals (which we cannot yet do), I would expect that we would find that some humans today are more closely related to a Neanderthal in any particular nuclear gene under study than to another human. Clearly, seen over the whole genome, Neanderthals were therefore not that different from us.

Another interesting aspect is that we are now beginning to get information about the genetic variation among Neanderthals and can start comparing it to humans and apes. A major difference between modern humans and modern great apes (chimps and gorillas) is that humans have relatively little genetic variation, whereas the great apes have significantly more. This raises the question as to whether the Neanderthals had little genetic variation like modern humans, or a lot, like the modern great apes. To date, there are five Neanderthals

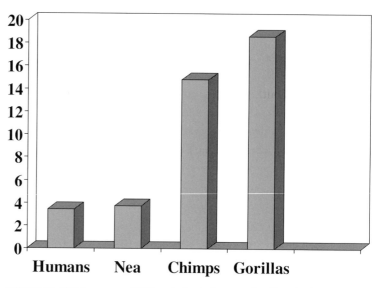

FIGURE 3.9 Within-group mtDNA variation in humans, five Neanderthals (Nea), chimpanzees and gorillas. Neanderthals seem to be similar to humans in having less variation than the great apes.

for which mitochondrial DNA sequences have been reconstructed. In addition to the type specimen, a Neanderthal from the Caucasus has been studied by Scottish and Russian scientists, and we have studied two individuals from Croatia and a second individual from the Neanderthal site. Of course, five individuals are not enough to provide a clear answer to that question but they provide a starting point. We found that the five Neanderthal individuals look very much like the modern humans in this respect (Figure 3.9). From this, I speculate that the Neanderthals were like humans in having relatively little genetic variation rather than like the apes in having a lot. We will be able to answer this question more reliably when we have more Neanderthal DNA sequences. It will then become possible to date the putative population expansion of Neanderthal ancestors out of Africa. This presumably happened earlier than the expansion to which contemporary humans trace their ancestry.

Maize domestication

A major revolution in the history of humans was the transition from a hunting–gathering lifestyle to an agricultural lifestyle. Central to this transition is the domestication of plants. I will discuss below what we are beginning to learn from the study of ancient DNA about when early farmers in the New World selected particular features during and after the domestication of maize.

Maize was domesticated in Mexico about 6000 years ago and then spread southwards to South America and northwards to northern Mexico and eventually into North America. We know that maize was domesticated from a grass, teosinte, to which a number of morphological and other differences were induced by early farmers through selection. For example, teosinte has major side-branches, which are suppressed in maize, and there is only one major corncob on the main stem of maize, whereas there are many little coblets on the side branches of teosinte.

We began analysis of maize some years ago using ancient corncobs collected from an area of South America where no teosinte grows. Therefore, maize must have been introduced here. The cobs varied in age from 400 years to 4000 years. Our first question concerned the genetic variation found in this early maize. We reasoned that if maize originated from one small spot, we would see a lot less variation in maize than in teosinte. The DNA was very well preserved in these plant remains, so we were able to study nuclear genes, specifically the gene coding for alcohol dehydrogenase. We found that ancient maize DNA sequences were often close to modern teosinte sequences and very far from other ancient maize sequences, which in turn were close to other modern teosinte sequences.

These data, which were obtained for a gene that probably was not selected by early farmers, contradicted our original naïve assumption, namely that all the maize would be closely related and teosinte would have a lot more variation. However, on more reflection, it is a useful situation. The reason is that, since most of the maize genome has as much variation as teosinte, one might expect reduced variation in maize to be hallmarks of genes encoding specifically selected traits. A fascinating recent development stems from the work of John Doebley and others, who have mapped a number of genes that are responsible for some of the traits that differ between maize and teosinte and found precisely this situation. In collaboration with Doebley and others, we have recently begun to study three of these genes. The first gene, 'teosinte branch 1', (tb1) is involved in the repression of the side branches found in teosinte, and makes maize have only one major stem. It is also involved in the formation of the singular cob on the major branch. The other two genes are involved in the protein composition and the content, structure and properties of starch in the kernels. The *tb1* gene is particularly well characterised, which allowed Doebley to compare the amount of variation in maize and teosinte at high resolution along the gene. One can divide the gene into two areas, one showing no difference in variation between maize and teosinte, and another area showing significantly less variation in maize than in teosinte. A phylogenetic tree reconstructed from the area of *tb1* showing no difference in variation between maize and teosinte provided the same picture we had seen before with the gene for alcohol dehydrogenase. However, in the region where maize has little variation and teosinte has a lot, the maize sequences all fall together in a tree and are intermingled only with a few teosinte types from the geographical area where maize was originally domesticated.

We have therefore chosen to study this part of the gene in ancient maize. In modern maize, there is only one type of DNA sequence while there are a lot of different sequences in modern teosinte. We have analysed samples from three 4000-year-old Mexican corncobs and from four corncobs found in New Mexico that vary in age between 1000 and 2000 years. In these ancient maize samples, we have found exactly the sequence that is found in contemporary maize, even if we go back to 4000 years ago. When we do the same analysis for the other genes, we find the same situation except in one case where a 2000-year-old samples carries a particular variant only seen in modern teosinte, but not in modern maize.

From these rather few samples we can already derive three insights: firstly, when going back in the domestication history as much as 4000 years, most of the selection has already taken place. Secondly, the selection may not have been entirely complete, however, because for one of the genes studied we still see a teosinte allele around about 2000 years ago. The third insight relates to the roles of these particular genes. One of them, *tb1*, is involved in the obvious structural difference of the plant in maize and in teosinte. Many people have argued that such differences were selected for early in domestication because they simplify harvesting the crop. However, the other two alleles that were already selected 4000 years ago contribute to the nutritional component of the kernel, the starch component and the protein content. Thus, early farmers selected not only obvious external traits of the plant that had to do with how easily corn can be harvested, but also biochemical traits that may have had to do with, for example, the pasting properties of the starch. By analysing more samples we will hopefully be able to get a more complete picture of what features were selected and at what time.

A further note of interest concerns the use of the maize by early farmers. Today, most maize is used to feed animals. Was the ancient maize also used for animals, or was it eaten by humans? In some cases, we can actually answer this question because we have matching remains of humans from the sites where the maize was found. These human remains are coprolites (i.e. old faeces) found in the same caves as the 2000-year-old corncobs. After extracting DNA from these coprolites, we were able to detect a particular maize allele in this sample. Therefore, we know that the maize carrying this particular DNA sequence variant was actually ingested by humans 2000 years ago.

Future prospects

In what areas of the study of ancient DNA may we expect to see progress in the next five years? Although it is hard to predict the future, there are two areas that I would like to single out. The first area is the retrieval of nuclear DNA sequences from Pleistocene organisms. This would open the prospect of studying genes involved in interesting features of these organisms. The second area is the retrieval of DNA sequences from many individuals within Pleistocene species so that population changes before and after the last Ice Age can be monitored.

Nuclear DNA from the Pleistocene

As discussed for the domestication of maize, we have been able to retrieve single copy nuclear genes from 4000-year-old samples. We have also been able to retrieve 40 000-year-old nuclear DNA from mammoth remains found in the permafrost, but so far not from non-frozen remains of that age. Therefore, one area of future research is trying to retrieve nuclear DNA from other Pleistocene remains. We focus on coprolites of extinct animals that are found in many caves in the south-west of the USA. For example, there are many caves that giant ground sloths have used as toilets over thousands of years, leaving layers of big remains behind from which we can now extract DNA. We have recently been able to amplify fragments of a nuclear-encoded blood clot factor gene from these samples that are up to 130 base pairs long. Therefore, we can now retrieve nuclear DNA from this extinct creature, string several pieces of DNA together and see, with great resolution, how it is related to living modern relatives. Still an open question is the extent to which we can do this from other remains. It would be highly interesting, for example, to be able to work with nuclear DNA from Neanderthals to generate a more complete picture of the Neanderthal gene pool. We would then be able to analyse genes in the Neanderthals that are involved in particular traits of interest. For example, a recently identified gene that is involved in speech and language would be an obvious choice. However, before that can be achieved, not only the technical retrieval of nuclear DNA from the Pleistocene must become possible. We must also develop some hitherto unknown way to distinguish such DNA sequences from contaminations from contemporary humans.

Pleistocene population dynamics

Another interesting direction is the retrieval of DNA from many individuals of Pleistocene species to reconstruct their population history and gain insights into their palaeodemography and palaeoecology. Currently, our lab is concentrating on cave bears, large herbivorous bears that existed in Europe and Western Asia in almost the same area as Neanderthals. We and others have by now retrieved DNA sequences from over 100 individual bears. In a phylogenetic tree of cave bear and contemporary brown bear mitochondrial DNA, all cave bears fall within a branch outside contemporary brown bears, just as Neanderthals fall outside contemporary modern humans. Even more interesting is that we can begin to see how different cave bears are related to each other across Europe before the last Ice Age. One striking observation is that, within many populations, there is great continuity over time. For instance, in two caves which are located about 10 km away from each other in the Alps, the same mitochondrial type existed in one cave from around 50 000 to around 30 000 years ago, while another type existed continuously over the same 20 000 years in the other cave. After this time, there are no more bears in these caves. In two other caves we found another interesting pattern: in one of the caves, there was continuity over 20 000 years, and in the other there was continuity over 10 000–12 000 years until a sudden replacement with a different type occurred.

By studying many Pleistocene cave bears from many locations, we are thus beginning to see interesting patterns. Alan Cooper's lab in Oxford is doing this for several other late Pleistocene animals. This will allow us to reconstruct the genetics of the palaeo-fauna before, during and after the last glaciation. Such knowledge can then be related to climatic change, the disappearance of the Neanderthals, the appearance of the modern human and other interesting phenomena of the time.

FURTHER READING

S. Pääbo, 'Ancient DNA', *Scientific American* **269**(5), 1993, 86–92.

M. Krings, A. Stone, R. W. Schmitz, H. Krainitzki, M. Stoneking and S. Pääbo, 'Neandertal DNA sequences and the origin of modern humans', *Cell* **90**, 1997, 19–30.

M. Hofreiter, D. Serre, H. N. Poinar, M. Kuch, and S. Pääbo, 'Ancient DNA', *Nature Reviews: Genetics* **2**, 2001, 353–60.

4 DNA and cancer

RON LASKEY

MRC Cancer Cell Unit, Hutchison/MRC Research Centre, Cambridge

DNA lies at the heart of our understanding of cancer and its treatment. Firstly cancer is essentially a disease of damaged DNA. It arises in a single cell that accumulates multiple genetic changes caused by damage to its genetic material, DNA. These disrupt the control mechanisms that regulate cell growth and turnover, allowing unrestrained cell division and tumour growth. Secondly and paradoxically, two of the most important treatments for cancer are based on inflicting yet more damage on DNA. Thus radiotherapy and many of the drugs used for chemotherapy attack cancer cells by breaking DNA to such an extent that the damaged cells commit suicide. The principle is one of fighting fire with fire, or specifically of fighting DNA damage with more DNA damage. This essay will consider the central relevance of DNA to both the cause and the treatment of cancer. In the final section, a further connection between cancer and DNA will be considered, namely how our increasing knowledge of DNA and its synthesis can be exploited to improve diagnosis and treatment of a wide range of cancers.

Cancer will affect one in three of the UK population and at current rates it will cause the death of one in four of them. It is tempting to ask, therefore, why cancer is so common. However, Gerard Evan has pointed out previously that a more appropriate question is: 'Why is cancer so rare?' As the human body contains more than 100 million million cells, and as cancer arises from breakdown of the rules controlling cell population sizes, it is remarkable that these rules fail so rarely. There are a thousand times more cells in the human body than the number of stars in our galaxy, the Milky Way, or 16 000 times more cells in each of our bodies than the number of people on the planet.

Clearly less than one in 100 million million cells successfully evades the rules that control cell proliferation. If antisocial, selfish behaviour amongst humans was as rare as it is amongst our cells, the prison overcrowding problem would be a thing of the past.

These figures become even more startling when they are viewed in the light of genetic stability. Before a cell can divide, it must duplicate its genetic material, DNA. Errors in that process, or failure to correct changes that arise from DNA damage, can result in the genetic changes that contribute to cancer.

The cell division cycle and programmed cell death

When human cells are removed from the body and cultured in nutrient medium on a plastic surface, they are able to divide every 24 hours. If all the cells in the body divided at this rate, we would double in size every 24 hours and be spherical. Clearly, patterns of cell proliferation in the body are tightly regulated and restrained. The fundamental process of cell division is part of a cell division cycle in which division alternates with synthesis of DNA. Before a cell can divide it must double its genetic content, DNA, to produce two identical copies for distribution between the two progeny cells. In addition to these discrete phases of DNA synthesis and cell division, there are two gaps, G1 which follows division and precedes DNA synthesis, and G2 which follows DNA synthesis and precedes cell division. During exponential growth these four phases follow each other cyclically: G1, DNA synthesis, G2, cell division, etc. (Figure 4.1). However, in the body, exponential growth is prevented by two additional levels of regulation. First, most cells in the mature adult body have escaped from the cell division cycle into a quiescent resting phase. In this phase they are neither synthesising DNA nor dividing. A sophisticated regulatory network of control circuits determines how many cells are allowed to escape from this phase to divide. The components of these circuits will be discussed below in the section entitled 'Patterns of cell proliferation are regulated by genes called oncogenes and tumour suppressor genes'. The second level of regulation that restrains cell population sizes is programmed cell death. Very few cells of the body are immortal and these are called 'stem cells'. Instead, the vast majority of cells are pre-programmed to undergo a dramatic and irrevocable suicide. Most of our cells are programmed to live for only a few weeks or even days. Immortality,

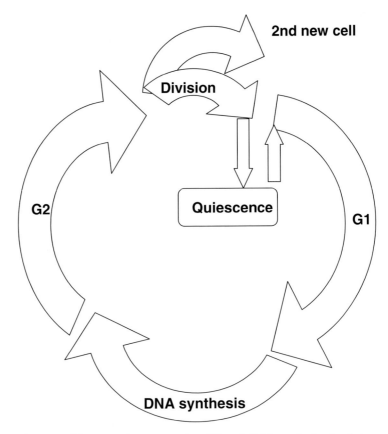

FIGURE 4.1 Diagrammatic representation of the cell division cycle showing that DNA synthesis occurs during a discrete phase of the cell cycle, preceded and followed by gaps G1 and G2, respectively. After gap G2 the cell divides and each new cell enters gap G1 before synthesising DNA again. Alternatively, after division, cells may enter a prolonged quiescence period, also known as G0. Most cells in the body are not proceeding around the cell division cycle but are arrested in the non-proliferating quiescent state.

that is survival for the entire lifetime of the organism, is the rare privilege of stem cells.

The process of cell death is extraordinarily decisive. Once a signal has activated the cell death programme, that signal is amplified via a cascade of enzymes, which culminate in the simultaneous mass destruction of essential

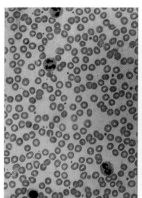 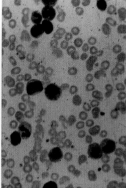 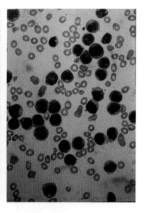

Normal blood

Chronic myeloid leukaemia

Acute myeloid leukaemia

CHAPTER 4 PLATE I
Blood smears from normal blood, chronic myeloid leukaemia and acute myeloid leukaemia. Compared to normal blood, chronic myeloid leukaemia contains an excess of mature and immature white blood cells circulating in the blood. In acute myeloid leukaemia, immature progenitor cells circulate in the blood in large numbers, instead of being confined to the bone marrow.

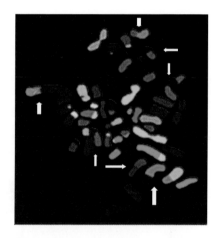

CHAPTER 4 PLATE II
Chromosome translocations seen in a breast cancer cell arising from breakage and incorrect rejoining of chromosome arms. Mutation of the breast cancer susceptibility gene, BRCA2, increases the probability of chromosomal rearrangements and hence of cancer (image provided by Paul Edwards).

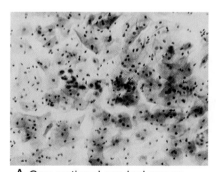 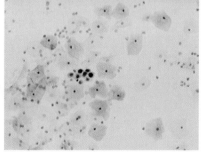

A Conventional cervical smear

B Same sample as A, but also stained with MCM antibody (brown)

CHAPTER 4 PLATE III Comparison of a conventional cervical smear stained by the Pap stain with a smear in which a light Pap stain is superimposed over an antibody stain for a novel marker (see text). Pre-malignant cells are stained brown against the background of pink and blue cells (reproduced, with permission, from Baldwin, Laskey and Coleman, *Nature Reviews: Cancer* **3**, 2003).

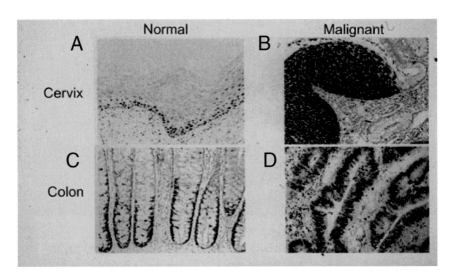

CHAPTER 4 PLATE IV Patterns of cell proliferation in normal versus malignant tissues of human cervix and colon visualised by staining proliferating nuclei brown with an antibody against a marker. The markers are called MCM proteins and are characteristic of cells which are anywhere in the cell division cycle except quiescence (see Figure 4.1). Proliferating nuclei (stained brown) are seen in discrete layers in normal cervix or normal colon. Cells at the surface of each of these tissues lack a marker. In contrast, cells containing the marker are found at the surface of malignant and pre-malignant tissues (brown nuclei in B and D).

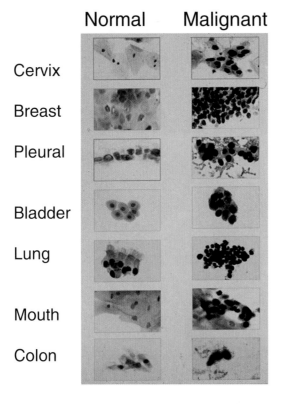

CHAPTER 4 PLATE V Comparison of normal versus malignant cells recovered from body fluids and stained with antibodies against the MCM cell proliferation marker. The left-hand column shows cells recovered from body fluids of normal volunteers, whereas the right-hand column shows cells recovered from patients with malignant or pre-malignant diseases of the equivalent site (brown nuclei).

proteins as well as of DNA. The cell's DNA is cut into over a million small pieces. Similarly many of its most essential proteins are simultaneously destroyed. Together these processes ensure that cell suicide is decisive and complete, once activated. The importance of this to the organism is clear. Incomplete cell death would allow damaged cells to persist and, as explained below, this would greatly increase the probability of cancer arising.

The extent of cell death in an organism is extraordinary. In the time it takes to read this chapter, several billion of each reader's cells will have died, thankfully not all neurones. Fortunately, in the same time several billion more cells will have arisen by cell division to replace those killed by programmed cell death – perhaps unfortunately, not all neurones. Remarkably the process of cell division and that of cell death are precisely balanced to maintain populations at a steady state, yet to allow wound healing, repair and limited amounts of tissue regeneration.

The immortal stem cell and its role in cell renewal

The restriction of immortality to a small minority of cells, the stem cells, provides a level of defence against cancer. Most cells can survive for only a limited time and divide only a limited number of times before cell death. Therefore, even if they start dividing at the wrong time or in the wrong place, they can only do so for a limited time before they die, eliminating the possibility of forming a tumour. For a tumour to form, its cells must also become immortalised (see below). Immortal stem cells divide only infrequently, producing one mortal 'committed progenitor cell' and one replacement stem cell. They self-renew. Progenitor cells divide for a limited number of times to expand the pool of cells arising from each stem cell division (Figure 4.2). Progenitor cells and their cell progeny are committed to a specific pathway of cell specialisation and to a specific cell lifetime. Their division produces the range of mature specialised cells that make up the tissues of the body. Tissue architecture arises from tightly regulated patterns and positions of dividing cells. For example, the surface layer of the skin, called the epidermis, is generated from a single layer of cells at its base, called the basal layer. This contains the stem cells and some of the committed progenitor cells that arise from them. As cells mature, they flow progressively from the basal layer to the superficial surface layers, where they die and are shed.

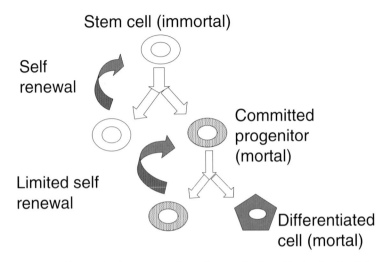

Stem cell (immortal)

Self
renewal

Committed
progenitor
(mortal)

Limited self
renewal

Differentiated
cell (mortal)

FIGURE 4.2 Immortality is a rare privilege of stem cells. A small minority of cells in the body are the stem cells that serve as a reservoir for replacing worn-out or damaged cells. Stem cells are immortal in the sense that they persist for the entire lifetime of the individual. When stem cells divide, they either produce two stem cells or one stem cell and one committed progenitor cell. Further division of the committed progenitor cell expands the pool of new cells and produces differentiated, specialised cell products. Unlike stem cells which have unlimited powers of self-renewal, committed progenitors have limited self-renewal capacity allowing a limited number of divisions before all of their progeny differentiate and subsequently die.

Patterns of cell proliferation are regulated by genes called oncogenes and tumour suppressor genes

Cell proliferation is tightly regulated by two classes of genes. It is switched on by genes called oncogenes, which encode a network of controls that transmit signals between cells and from the cell surface into the cell nucleus. These signals switch on DNA synthesis, which leads in turn to cell division. In contrast to oncogenes, which control cell proliferation by switching it on, tumour suppressor genes control cell proliferation by switching it off. They provide safeguards to ensure that cells only proliferate in the right place at the right time. They can be regarded as the equivalent of the handbrake of a car. Only when the handbrake is released can the car move forward. Similarly cells can only respond to proliferation signals from oncogenes after they have been released from the inhibition by tumour suppressor genes. This dual level of controls, switching on by oncogenes and switching off by tumour suppressor genes, allows a robust, and normally fail-safe, mechanism to prevent cells from proliferating in the wrong place at the wrong time.

How do oncogenes switch cell proliferation on?

There are many levels at which the products of oncogenes do this. Some oncogene products are growth factors which carry signals from one cell to another in much the same way that hormones signal between different cells of the body. Other oncogene products serve as receptors which detect growth factors at the cell surface and pass the signal on to the inside of the cell to report that a stimulatory growth factor has arrived. Yet more oncogene products provide a finely balanced network of signal relay, integration and amplification systems that result in activating the molecular machinery of DNA synthesis and cell proliferation. Interestingly, many of these signalling molecules were discovered because human tumours were found to contain changes in the genes that specify their production. These genes are mutated or damaged so frequently in human tumours that this fact led to their discovery. Examples of oncogene changes in cancer are given in Table 4.1. These findings have encouraged further systematic studies of the DNA of cancer cells in order to identify which changes are important in any particular kind of cancer. A recent triumph of this approach has been the discovery that one signalling molecule called B-Raf is present in too many copies in a wide range of malignant melanomas. This study was performed by Mike Stratton and colleagues at the Sanger Centre,

Table 4.1. *Examples of oncogenes mutated in cancers*

Name	Type of mutation	Tumour
Ras	Point mutation	Many carcinomas, melanomas, etc.
Myc	Amplification	Many, including breast, cervix, brain, lung, etc.
Abl	Translocation	Chronic myeloid leukaemia
Raf	Amplification	Melanoma
Neu	Amplification	Breast
Neu	Point mutation	Neuroblastomas

named after Fred Sanger who won his second Nobel Prize for inventing methods of determining the sequence of information in DNA. Now that this study
at the Sanger Centre has identified B-Raf as a key player in the formation of
malignant melanoma, it encourages a fresh attack on that disease by seeking
new drugs which can specifically inhibit B-Raf itself and hopefully combat one
of the most aggressive and troublesome of human tumours.

Tumour suppressor genes serve as safety checks for cells which are preparing to proliferate before they can divide or make DNA

Like oncogenes, tumour suppressor genes can act at many different levels. Some
of them directly counteract the signal relay and amplification systems produced
from oncogenes. An example is the retinoblastoma or Rb gene. The balance of
activities between this gene product and the stimulatory oncogene signalling
pathways determines whether a cell will divide or not. An alternative level
of action of a tumour suppressor gene is demonstrated by a gene that is very
frequently mutated in a wide range of human cancers and is called p53. It
performs the equivalent role of a fire alarm. It responds to damage by calling a
halt to cell proliferation until the damage has been repaired. More specifically,
it senses damage to DNA and ensures that that damage is not copied into new
DNA, which would result in a change to the genetic information of the progeny
cells. Thus it either causes a pause in DNA synthesis to allow the damage to
be repaired or, if the damage is too severe, it triggers programmed cell death

Table 4.2. *Examples of tumour suppressor genes that protect against cancer*

Gene name	Function in the cell	Inherited cancers associated with a mutated gene
Rb	Cell cycle checkpoint	Retinoblastoma
p53	Detects and signals DNA damage	Many. Li-Fraumeni syndrome
APC	Signalling within the cell	Colorectal cancer
BRCA1 & 2	Maintain genetic stability	Breast cancer
p16	Cell cycle checkpoint	Melanoma
VHL	Regulates breakdown of certain proteins	Kidney and brain

of the damaged cell. Thus it serves as an editing mechanism to stabilise the genetic content of the cell, preventing the production of cells with an altered DNA content, which could rapidly progress to form tumours. Mutation of p53 in tumours accelerates the accumulation of genetic changes contributing to tumour progression. Examples of changes in tumour suppressor genes in cancer are given in Table 4.2.

Carcinogens cause cancer by mutating the DNA of oncogenes and tumour suppressor genes

It is widely known that environmental agents such as tobacco, asbestos or ionising radiation cause cancer. So how do these relate to the oncogenes and tumour suppressor genes whose damage has been blamed for cancer in the previous section? The answer is relatively simple. Carcinogens such as tobacco smoke cause mutations in DNA and when these occur in the DNA of oncogenes or tumour suppressor genes, they result in cancer. Thus DNA is the target of carcinogens. Carcinogens can cause many different types of change in DNA and some of these are illustrated in Figure 4.3. They range from simple changes of one letter in the DNA sequence on the one hand, to complete deletion or duplication of a region of DNA. For example, many human tumours contain a mutation of a specific single letter of the DNA, which codes for a small oncogenic regulatory protein called Ras. This change of a single amino acid building block of the protein Ras makes it become hyperactive, triggering repeated cell proliferation. Alternatively, deletion of a tumour suppressor gene or amplification of an oncogene both occur in many human tumours. Examples include the tumour suppressor genes BRCA1 and 2, which are frequently mutated or deleted in breast or ovarian cancers, whereas tumour suppressor genes p53 and Rb (retinoblastoma) are frequently mutated or deleted in a much wider range of tumours. There are many examples of oncogene amplification resulting in excess copies of the oncogene. One example is the amplification of the oncogene Myb in some cases of acute myeloid leukaemia (Plate I), though other genetic changes also contribute to this form of leukaemia, such as the type of change described in the next paragraph.

Another extremely important change in the DNA of cancer cells is called chromosome translocation. It involves breakage of DNA in a chromosome, followed by subsequent aberrant rejoining. Instead of rejoining to the original

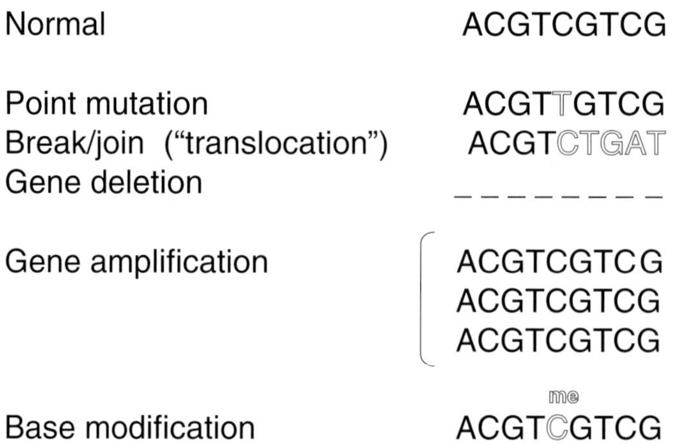

Normal	ACGTCGTCG
Point mutation	ACGTTGTCG
Break/join ("translocation")	ACGTCTGAT
Gene deletion	– – – – – – – –
Gene amplification	ACGTCGTCG
	ACGTCGTCG
	ACGTCGTCG
	me
Base modification	ACGTCGTCG

FIGURE 4.3 Carcinogens can damage DNA in several ways, depicted in this diagram. They can change a single letter in the genetic script (point mutation), cause breakage followed by incorrect rejoining (chromosome translocation), delete the entire sequence of a gene or part of that gene, amplify a gene or multiple genes so that there are many copies of the sequence encoding a particular gene product, or cause base modifications such as methylation of the nucleoside cytosine, which has the effect of silencing a gene even though it is still present (diagram modified, with permission from Mike Stratton).

partner, one severed chromosome end joins to part of another chromosome, bringing two distant regions of DNA together. This can either disrupt the correct regulation of a gene situated at the boundary, or it can result in synthesis of a new fusion protein, containing components of two different original proteins fused together. One of the best known examples of this contributes to the cause of chronic myeloid leukaemia (Plate I). It involves fusion of two genes called Bcr and Abl, bringing segments of chromosomes 9 and 22 together in one fused abnormal chromosome. The function of the Abl oncogene is known. It attaches phosphates to certain other proteins, altering their activities. This has made it possible to screen for drugs which inhibit Abl's activity, resulting in discovery of the drug Glivec, which has dramatically improved the treatment of chronic myeloid leukaemia. This paradigm also brings hope that similar specific drugs may be developed to treat other cancers in which a specific oncogene is found to be activated. Chromosome translocations are relatively common in tumours related to the immune system, such as leukaemias and lymphomas. This appears to be due in part to the fact that antibody diversity is generated by deliberately breaking and reshuffling pieces of DNA to establish the enormous repertoire of different antibody specificities. If breakage is followed by rejoining to an incorrect partner from another chromosome, this can produce the carcinogenic consequences outlined above. Although chromosome translocations are particularly common in tumours arising from white blood cells, chromosome translocations are also found in many solid tumours (Plate II). Their significance has been debatable, though now that highly specific reagents to identify chromosome regions have become available, evidence is growing that chromosome translocations may play an important part in a wide range of other tumour types as well.

In addition to mutation, gene deletion, gene amplification or chromosome translocation, one specific modification of DNA is found in many tumours. It is DNA methylation and it involves addition of a methyl group to the base cytosine at intervals along the DNA. This occurs on a subset of cytosine–guanine sequences and it has two consequences. First, it silences the adjoining genes, so even though they have not been deleted or mutated, they remain firmly switched off. Second, the sequence cytosine–guanine forms base pairs with the sequence cytosine–guanine on the opposite 'antiparallel' strand (cytosine forms base pairs with guanine and vice versa). Therefore methylation is a heritable modification. Methylation on one strand is copied to the opposite strand after replication and the pattern of gene inactivity specified by DNA methylation is passed on to progeny cells. Hence DNA methylation of tumour suppressor genes is frequently found in human tumours, where it allows them to be switched off, allowing the cancer cell to escape from cell proliferation controls.

There is another DNA change which plays an important part in a minority of cancers, namely infection by a tumour virus. This is the underlying cause of primary liver cancer, cervical cancer and a small number of other cancers such as Burkitt's lymphoma. Hepatitis B and C viruses cause primary liver cancer, whereas cervical cancer is caused by papilloma viruses, which infect cells near the surface of the cervical tissue and stimulate their proliferation. The DNA of papilloma viruses codes for two proteins that selectively target and inactivate two tumour suppressor genes, Rb and p53 (see above), resulting in excessive cell proliferation.

Cancer is a multi-step process

Cancer arises from the progressive accumulation of DNA damage arising from carcinogenic attack. Occasionally the damage will occur in the DNA of oncogenes or tumour suppressor genes. If these events increase the activity of oncogene products or decrease the activity of tumour suppressor gene products, then the affected cell can escape from the regulatory mechanisms restraining cell population growth (Figure 4.4). What follows is a Darwinian selection process. Cells which have randomly activated oncogenes or inactivated tumour suppressor genes will continue to divide and will increase in number even though the population of normal cells stays constant. Programmed cell death would normally limit the impact of this unrestrained cell division by limiting the lifetime of the cell and its progeny. However, if a second mutation occurs in the genes that control cell survival versus cell death, then the rapidly dividing cells with mutated oncogenes and tumour suppressor genes can persist to form a tumour. This in itself need not be lethal. A well-localised tumour that is restricted to a single location can be removed by surgery or ablated by localised radiotherapy. However, tumours become much more difficult to treat if they have also accumulated mutations in the genes that regulate cell movement and migration. Increased movement and migration can lead to invasiveness so that cells break away from the site of the original primary tumour and form secondary tumours dispersed throughout the body. This process is called metastasis and it makes cancer much more difficult to treat. Local therapies become ineffective once cells have metastasised to remote parts of the body and established secondary tumours there. Once metastasis has occurred, chemotherapy is required to attack cancer cells wherever they may be in the body, rather than just at the localised site of the primary tumour.

- Transformation: cells evade controls
- Immortalisation: cells evade death
- Invasion: escape to colonise new sites
- Angiogenesis: attract new blood vessels

FIGURE 4.4   Tumour formation occurs in multiple steps.

In addition to mutations in oncogenes, tumour suppressor genes, cell death genes and genes that regulate cell migration, tumour growth can be boosted still further by mutations in genes that regulate blood vessel formation. Without a dedicated blood supply, tumour growth will be limited, but that restraint can be lifted if the tumour secretes growth factors to attract new blood vessels to invade the tumour, bringing nutrients and oxygen. Maximum tumour growth depends on the combination of each of these multiple steps in tumour formation. This is one of the reasons why the incidence of cancer is extraordinarily low compared to the number of cells in the body. Figure 4.5 attempts to illustrate the multi-step nature of cancer by analogy with the board game 'Snakes and ladders'. This analogy is not attempting to trivialise the seriousness of the disease, but it provides a simple conceptual framework for looking at the effects of the multiple steps in tumour formation from the dangerous perspective of the tumour cell. Progress from a normal cell (bottom left) to a life-threatening tumour (top left) can be made in very many small steps or dramatically accelerated by the major events depicted by the ladders in the cartoon. The events illustrated include inactivation of tumour suppressor genes, inactivation of oncogenes and immortalisation by an inactivation

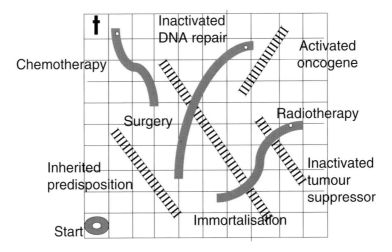

FIGURE 4.5 Progression of a normal cell to a cancer cell and a malignant tumour can be accelerated by the specific events depicted as ladders in this sinister cartoon which is based on the board game 'Snakes and ladders'. The significance of these events is explained in the text.

of cell death genes. However, two other circumstances depicted by ladders are fundamentally important and these form the subject of the following section.

Inherited predispositions to cancer

Studies of the incidence of particular types of cancer amongst families have accumulated evidence that certain families are predisposed to particular types of tumour. Two of the best-known examples are inherited tendencies towards breast cancer and inherited tendencies towards colon cancer, though many other examples exist. One of the first examples to be understood clearly at a molecular level was the inherited predisposition to cancers of the retina of the eye, retinoblastomas. Children of a parent with retinoblastoma have close to a 50% chance of developing the disease themselves. This is because they have inherited one normal copy and one damaged copy of the retinoblastoma tumour suppressor gene Rb. This gene prevents tumours as outlined above. Normal cells have two copies of the retinoblastoma gene, one from each parent. Either of these two genes can protect cells from inappropriate cell proliferation. Therefore damage to one of the two by a random mutagenic event is relatively unimportant as the second gene can still restrain the cell from dividing. The probability of a second retinoblastoma gene becoming damaged during the lifetime of a cell with one damaged retinoblastoma gene is vanishingly small. However, inheritance of one damaged retinoblastoma gene from one parent means that there is only one functional gene left to defend against excessive proliferation, so that a single mutagenic damage event is sufficient to remove the restraint on cell proliferation. A similar situation applies in the case of many other inherited predispositions to cancer. For example many, but not all, predispositions to breast cancer arise from inheritance of a damaged BRCA gene. The probability of random damage inactivating the second gene is very many times higher than the probability of inactivating two normal genes in the same cell. Hence progress through the multiple steps of carcinogenesis illustrated in the cartoon in Figure 4.5 is dramatically accelerated.

This acceleration effect can be particularly marked when the damaged gene is one that is responsible for repairing DNA damage, as this damage can lead to rapidly increasing genetic instability with concomitant acceleration of tumour progression.

Paradoxically DNA damage is used extensively in cancer treatment

Although cancer is caused by DNA damage, it is ironic that two of the main approaches to cancer treatment are based on inflicting yet more damage on DNA (Figure 4.6). The three most commonly used treatments in cancer therapy are surgery, radiotherapy and chemotherapy. Although surgery is not based on DNA, radiotherapy and many of the drugs used in chemotherapy owe their effects to the damage that they cause to it. Specifically, many of them are designed to break DNA. Thus the focused beams of ionising radiation used for cancer radiotherapy cause breaks across both strands of DNA. Drugs used for chemotherapy act in many different ways, but several break either one or both strands of DNA. Others cross-link the two strands together or insert into DNA in such a way that they distort the double helix. Examples are given in Table 4.3. The use of DNA-damaging agents to treat cancer raises an obvious paradox. If cancer arises from damage to DNA, how can damaging DNA still further cure cancer? The answer is that too many DNA breaks trigger cell death. Therefore the aim is to inflict so much damage selectively on the cancer cells that they die leaving the normal cells unaffected. In the course of radiotherapy this is achieved by focusing radiation so that a higher dose is delivered to the tumour than to the surrounding tissues. In the case of chemotherapy, cancer cells are more vulnerable for two reasons. First, they never leave the division

SURGERY   Not DNA based (but vital!)

RADIOTHERAPY   Radiation: breaks DNA

CHEMOTHERAPY   Drugs: many break DNA

1. Too many DNA breaks trigger cell death
2. Cancer cells are more vulnerable because
   a. They never leave the division cycle
   b. They ignore danger signs (checkpoints)

FIGURE 4.6 Paradoxically DNA damage is the basis of radiotherapy and many chemotherapy drugs.

Table 4.3. *Examples of DNA-damaging drugs used in chemotherapy*

Cisplatin	Binds DNA
Etoposide ⎱ Doxorubicin ⎰	Prevent DNA joining after natural breakage
Methotrexate ⎱ Fluorouracil ⎰	Inhibit DNA synthesis by depleting precursors
Bleomycin ⎱ Mitomycin ⎰	React with DNA, breaking or cross-linking it

cycle, unlike normal cells which spend only part of their time in the division cycle. Thus they never go to ground by becoming quiescent. Second, cancer cells ignore the danger signs, the checkpoints that arrest proliferation in normal cells until damage is repaired. Characteristically, these checkpoints are frequently silenced in cancer cells so that the cell continues to make DNA even if the DNA is damaged. This increases the risk of catastrophic cell death of the cancer cell in response to the damage inflicted by the drug. Extensive effort is currently focused on improving the selectivity of both radiotherapy and chemotherapy. For example, if radio-sensitising agents can be incorporated into the tumours selectively then the differential effect of radiotherapy on the tumour compared to the surrounding tissues could be enhanced. Similarly methods of delivering drugs selectively to tumour tissue are under investigation. Once again the aim is to maximise damage to the tumour cell and minimise damage to the normal cell.

Improving early detection of cancer using DNA synthesis proteins as markers

Localised tumours are much easier to treat than tumours which have spread around the body by invasion and metastasis. Therefore early detection of cancer has an important role to play in detecting tumours before they have evolved into invasive metastatic secondary tumours. This is the basis of screening programmes for cervical cancer and breast cancer. Deaths from cancer of the cervix of the uterus have been dramatically reduced by introduction of a regular programme of cervical smear tests. These sample the cells on the surface of the cervix, smear them on to a slide and stain them with dyes which attempt to highlight differences between normal cells and cancerous or pre-cancerous cells. Although this test is an extremely important public health measure, saving at least 1000 lives a year in Britain alone, yet it is notoriously difficult to interpret accurately. All too often errors of interpretation occur, which can have serious consequences for the patient. When errors occur, the media are very keen to apportion blame, but the fact is that this test requires extraordinary levels of training and concentration. It is inherently difficult to distinguish all abnormal cells from normal cells and this is made more difficult by the fact that abnormal cells will be a rare minority amongst normal screening samples. We have attempted to use knowledge of the biology of cell proliferation and cell organisation in the cervix to develop a test for pre-cancerous changes, which stains the abnormal cells a different colour from the normal cells, thus making them much easier to detect. Plate III shows the difficulty of detecting abnormal cells by the conventional Pap smear test, compared to the relative ease of detecting them when the Pap smear test is superimposed over staining for a particular marker, called an MCM protein. MCM stands for a relatively uninformative name in this context, mini-chromosome maintenance, which refers to the way in which these proteins were discovered in yeast cells. There are 6 MCM proteins and they form a ring shaped complex containing one each of the six. The MCM complex plays an essential role in DNA synthesis, probably unwinding the two strands of the double helix from around each other so they can be copied into two daughter helices. MCM proteins are abundant within the cell nucleus throughout the cell division cycle, but they are absent from cells which have become quiescent and stopped dividing. Significantly, most of the cells in the body are quiescent and are not in the division cycle. Proliferation is restricted to specific sites such as those where committed progenitors divide to expand the pool of new cells (Figure 4.2). Plate IV shows the patterns observed when sections through normal tissues are stained with an antibody raised against MCM proteins. Cells that contain MCM proteins are stained brown in their nuclei, whereas cells that lack MCM proteins are only stained blue. Panel A shows that cells containing MCM proteins (brown nuclei) are only observed in the basal layers of the cervix. This is where new cells form by division and they migrate from this site to the surface where they die and are shed. None of the cells at the surface show the brown MCM marker. In contrast, Panel B shows a section through a pre-malignant lesion of the cervix of exactly the type that a smear test is designed to detect. Brown nuclei containing the MCM marker are seen right up to the surface of the tissue and cells containing the marker are shed or sampled by the smear test, whereas this is not true of normal tissue. Panels C and D show that a similar difference is observed between sections of normal colon and colon carcinoma. In the case of the colon, cell proliferation (brown nuclei) is restricted to the bottom third of the deep crypts that line the large intestine (Panel C). Once again this is where new cells form before they progress to the cell surface to be shed into the gut. Note that the upper layers in normal tissue do not contain the MCM marker. In contrast the cells of a colon carcinoma stain brown with the MCM marker right up to the cell surface and cells containing the marker are shed into the gut (Panel D). Similar patterns have been observed for several other tissues. In practice these panels mean that an opportunity exists to screen for cancers by determining the presence or absence of the MCM marker in the cells that are shed into body fluids. This approach is illustrated for the MCM marker in cells of a cervical smear in Plate III. It is illustrated for cells recovered from a range of other sites in Plate V which illustrates the recovery of cells lacking the marker (blue nuclei) from the body fluids of normal volunteers in the left column and the presence of cells stained brown by the marker in cells recovered from the body fluids of patients with cancer or a pre-cancerous change. Clinical trials are in progress to assess the value of this approach for early detection of several of the commonest cancers, including cancer of the cervix, colon, lung, oesophagus, breast and bladder. There are bound to be technical hurdles to overcome before such a test can be adequately validated and made robust enough to function in routine clinical practice, but the trials to date suggest that the MCM markers have the potential to serve as screening tests for several of the commonest cancers.

DNA-based methods of tailoring cancer treatment to the specific molecular details of individual tumours

There are many different drugs for combating cancer by chemotherapy. The selection of the most effective drug or combination of drugs is vitally important for patient survival. However, there are many different oncogenes and tumour suppressor genes involved in controlling cell proliferation; and damage to different combinations of these can cause closely similar tumours which will respond differently to the same treatment. The more precisely the molecular defects of a particular tumour are understood, the more precisely the treatment can be tailored to that patient. Recent technological advances now make it possible to determine the genetic signature of an individual patient's tumours so that treatment can be matched as closely as possible to the underlying problems. There are two main ways in which this can be approached, both of them exploiting the ability to display a grid of specific DNA sequences in two dimensions and to interrogate these arrays of sequences using molecules present in the patient's tumour. One approach extracts DNA from the tumour and compares it to normal DNA to ask if oncogenes have been amplified or tumour suppressor genes have been deleted. The second approach does not look at the genes themselves, but at the messages that are copied from the genes in order to make specific proteins. So these are not looking directly at the DNA genome of the cancer cell, but they are looking at the amounts of specific gene products that are being made by the cancer cell. Hence they are called expression arrays. These approaches appear to be extremely promising and they are being actively evaluated for many types of cancer in many laboratories at present. For example, one of the most advanced studies has shown that breast cancer patients can be classified into those who have a high risk of invasion and metastasis and therefore poor outcome versus those who have a lower risk and better outcome, on the basis of the level of expression of 70 selected gene products. At the very least, these approaches will allow more informative clinical trials as drugs can be tested on tumours which are known to have a specific defect and this alone should allow the design of chemotherapy programmes which are more effective against specific tumours. These developments open an exciting new phase of opportunity to make substantial improvements in cancer therapy and they are firmly based on reading the information in the patient's DNA.

FURTHER READING

B. Alberts, A. Johnson, J. Lewis, M. Raff, K. Roberts, and P. Walter, *Molecular Biology of the Cell*, 4th edn, New York: Garland Science, 2002, ch. 23.

P. Baldwin, R. A. Laskey, and N. Coleman, 'Translational approaches to improving cervical screening', *Nature Reviews: Cancer* **3**, 2003, 217–26.

D. Hanahan and R. A. Weinberg, 'The hallmarks of cancer', *Cell* **100**, 2000, 57–70.

J. K. Heath, *Principles of Cell Proliferation*. Oxford: Blackwell Science Ltd, 2001.

H. Lodish, A. Berk, S. L. Zipursky, P. Matsudaira, D. Baltimore, and J. Darnell, *Molecular Cell Biology*, 4th edn, New York: W. H. Freeman and Company, 2000, ch. 24.

P. Nurse, 'A long twentieth century of the cell cycle and beyond', *Cell* **100**, 2000, 71–78.

R. A. J. Spence and P. G. Johnston (eds.), Oncology, Oxford: Oxford University Press, 2001.

B. Vogelstein, and K. W. Kinzler, (eds.), *The Genetic Basis of Human Cancer*, New York: McGraw-Hill Companies Inc., 1998.

5 DNA, biotechnology and society

MALCOLM GRANT

University College London

What is it that shapes social attitudes towards advances in biotechnology? Why have GM crops, now so widely grown in North America, Argentina and China, met with such resistance in Europe? The breadth of the issues, the divergence of the underlying values, public mistrust of Government and the polarisation of the debate within Europe, all suggest that the science–technology–society relationship is today far more complex than that identified by C. P. Snow in his famous 'two cultures' lecture in Cambridge in 1959. This essay will explore these issues against the dynamics of contemporary public debate over the potential commercialisation of GM crops in the UK.

Introduction

> At a few minutes past five o'clock in the afternoon of 7 May 1959, a bulky, shambling figure approached the lectern at the western end of the Senate House in Cambridge.[1]

The lecturer was the novelist, scientist and politician, C. P. Snow. In the course of the lecture that followed he kicked off a furious controversy that echoes around the world to this day. It was about what he called the 'two cultures'. He painted a caricature of a British intellectual elite divided between two camps: of scientists ignorant of the arts, and of literary intellectuals who could

---

[1] Stefan Collini, 'Introduction', to C. P. Snow's *The Two Cultures* (Cambridge: Cambridge University Press, 1998), p. vii. Collini comments that 'By the time he had sat down over an hour later, Snow had done at least three things: he had launched a phrase, perhaps even a concept, on an unstoppably successful international career; he had formulated a question (or, as it turned out, several questions) which any reflective observer of modern societies needs to address; and he had started a controversy which was to be remarkable for its scope, its duration, and, at least at times, its intensity.'

only be described as natural luddites. From his own painstaking observations at Cambridge college high tables, Snow had found that, if asked directly, the literary types were incapable even of describing the Second Law of Thermodynamics. Over the following months, this throwing down of the gauntlet to the literary luddites reaped its foreseeable reward, particularly within Cambridge. The charge was led by F. R. Leavis who mounted a particularly scurrilous personal attack on Snow's scholarship as well as his thesis.

Snow remained unrepentant. Interestingly, however, when he revisited his lecture four years later, he wondered whether he might have set the wrong test. He should, he thought, have asked that High Table question in terms not of thermodynamics, but of molecular biology. He had come to believe that:

> This branch of science is likely to affect the way in which *men think of themselves* more profoundly than any scientific advance since Darwin's – and probably more so than Darwin's.[2]

I expect that few today would disagree. Indeed, many may believe that the phenomenon of the two cultures is nowhere better demonstrated than in the current debate about DNA, biotechnology and society, particularly in the context of GM crops. In exchanges that are grotesquely reminiscent of Snow's lecture, those who oppose GM have been widely caricatured as luddites, ignorant of science, incapable of understanding the close similarities between GM and non-GM technologies for crop development and improvement, foolishly seeking a risk-free world and unconscionably obstructing the development of a new agriculture capable of feeding a growing and hungry world population.

A columnist in the *Observer*, lamenting the death of Dolly the sheep, spoke of 'our chattering classes [who] think GM foods are dangerous because they contain DNA, who believe horoscopes tell the truth, and who value media studies above an education in science'.[3] If only there were better scientific understanding on the part of the public, such critics seem to assume, our fields would already be blooming with GM crops and queues of joyful shoppers would be lining up to snatch GM food from supermarket shelves.

The caricature on the other side is equally stark, and is reinforced by a news media that occasionally falls short of the high standards of rigour and impartiality that we like to expect of the national press. Scientists, biotechnologists

---

[2] Snow, *The Two Cultures*, p. 74: emphasis in the original.
[3] Robin McKie, 'Goodbye Dolly – you leave the world a better place', *Observer*, Sunday 16 February 2003, p. 28.

and companies involved in GM technology are variously accused of scientific and technological determinism, of reckless arrogance and greed, of developing Frankenstein foods and creating super-weeds, of seeking world domination for privately developed and privately owned technologies, of seeking to interfere with nature with potentially irreversible consequences and of seeking to make a profit. Governments and politicians, led by the USA, are cited as conspirators wishing to promote biotechnology as the new major global growth industry to follow the revolutions in IT and communications.

These extreme camps are identified by Onora O'Neill in her essay in this book as, respectively, the 'enthusiasts' and, the 'scaremongers', and, following a thoughtful consideration of the ethical framework, she argues for regulation with care and rigour, in the light of the evidence.

Now, the rhetoric of the debate throughout Europe is deeply polarised, and resistant to any early resolution through conventional political machinery. I suspect that even C. P. Snow would have been amazed not only by the intensity and the bitterness of the exchanges, but also by the extent to which the parties, with their horns firmly locked, continue to speak right past each other.

How, though, does all of this play in people's real lives? Why have politicians had to become so concerned about the social embedding of new technologies? Or, to put it another way, what is it that 'society' really wants from new technology and from those who regulate it on their behalf? What are the values that underpin people's opinions? Must a lack of scientific understanding be taken to disable people from expressing informed preferences? And if people express informed preferences, ought they to be relevant to political choice and regulation? If so, how can they be captured and injected into those processes? Is the dispute over GM crops really to do with the technology as such, or to do with trust in those who own and control it? If the latter, what does this imply for any process of case-by-case evidence-based regulation?

Science and technology

Mankind has been developing and improving crops for thousands of years by selective breeding. Conventional plant breeding involves cross-pollination, growing and then selecting plants for particular traits. For the best part of the last century this process has been enhanced by scientific method. Mutations have been induced in seeds by chemicals or irradiation, bringing about wide-ranging and random genetic changes, from which plants with desirable traits can be selected. These processes have transformed crop improvement. There has been a growth in global agricultural production that has outstripped the growth in population: world output of food per head has gone up by some 25% over the past 40 or so years, during which time population has risen by some 90%.

Crick and Watson's discovery 50 years ago of the helical structure of DNA opened up a new chapter in our understanding of the genetic structure of plants. In parallel with the sequencing of the human genome has developed the sequencing of plant genomes, starting with the mouse-ear cress, or arabidopsis. This has a relatively small set of genes which determine how it functions as a plant, but it provides a valuable model for plants with much larger genomes such as corn, rice, wheat, soybean and cotton, all of which will no doubt be individually sequenced in due course.

Modern biotechnology relating to plants is developing across a number of different fronts. Two different aspects have assumed significance in the debate around GM crops.

(1) Marker-assisted breeding, in which a short sequence of DNA acts a tag for another, closely linked gene. This is effectively a genetic fingerprinting technique, which allows plant breeders to identify desirable traits based on genotype rather then phenotype, matching molecular profile to the physical properties of the variety. This allows significant acceleration in the speed of conventional plant breeding.

(2) Recombinant DNA technology, also referred to as genetic modification, genetic manipulation, genetic engineering and transgenesis, and abbreviated to 'GM'. In this process, DNA from any class of organism may be isolated and introduced to crop plants. It is more precise than conventional breeding, in the sense that the introduced genes are few in number and known. Moreover, the range of possible genes that may be inserted is much greater. Genes may come from another organism of the same species, perhaps replicating in the laboratory a process which could have occurred in nature. Or they may come from a wholly different species. A widely cited example is that of DNA being transferred from salmon to strawberries. It is the potential transformative capacity of the technology used in that way that has most excited public concern, especially when companies talk of designing wholly new plants for the future.

In relation to crops, the principal modifications that have so far led to commercial growing relate to insect resistance, in which a crop is modified to express its own insecticides, and herbicide resistance, which means that a crop may itself be sprayed with a broad-spectrum contact herbicide, such as glyphosate (Monsanto's Round-up) or glufosinate ammonium (marketed in the USA as Liberty Link), thereby enhancing weed control. This first generation of GM crops may have production benefits to seed producers and farmers, and may reduce chemical inputs to agriculture, but it has so far offered little of interest to European consumers. Other GM crops now in commercial production include cotton, soybeans, maize, oilseed rape (canola) and tobacco. Wheat and rice are close to commercial production. Forestry trees, forage grasses, vegetables and fruit have also been genetically modified and a number of these are in development, and some, for example papaya, tomatoes, peppers and melons, are in commercial production. Development is under way for technology to introduce or enhance such traits as the removal of allergens from peanuts, and to develop salt tolerance, drought tolerance, pharmaceuticals, nutraceuticals, biofuels, soil decontaminators and plastics. Further possibilities include the development of perennial rather than annual crops, enhanced nitrogen fixation and male sterility.

Adoption of GM crops

Internationally, GM crops have rapidly become a major phenomenon. In 1996 there were 1.5 million hectares of GM crops, primarily in the USA. By the year 2002 the area had risen to 58.7 million hectares, around 15 times the cropped area of Great Britain. The main countries are the USA, Canada, Argentina and China. The main crops are soybean (over half the world's production is now GM), canola (oil seed rape), maize and cotton. Around 75% by area is of crops modified for herbicide resistance, 17% for insecticide resistance, and the remaining 8% for both. GM food, either in live or in processed form, is now widely consumed around the world. It would be true to say that if there are food safety issues – a question which continues to be contested, not least by the British Medical Association (BMA) in evidence given last year to the Scottish Parliament – the epidemiologists are not short of raw material.

In Europe, however, the contrast could hardly be greater. No fresh GM produce has been cleared for sale in the UK. There is practically no commercial growing of GM crops: a few thousand hectares in Spain and Germany, and none at all in the UK. From the perspective of the US Government, whose farmers export around 30% of their produce, this is nothing less than a flagrant violation of the principles of free trade, embodied in the World Trade

Organisation (WTO). The Americans are similarly upset by the labelling and traceability proposals currently going through the EU law-making processes, which they see as a further barrier to free trade. There is nothing new about agricultural trade wars between the two blocs. Several years ago, the USA fought and won a similar action when EU officials banned US beef exports on the grounds that they contained unsafe growth hormones (BST). It was a lengthy process, however, and in the beef hormone dispute, the EU simply chose to pay a $100m fine each year rather than admit US beef products.

Underlying the EU position is concern about consumer opinion and its political counterpart, voter reaction. Despite an early flirtation with the FlavrSavr tomato, supermarkets in the UK cleared unprocessed GM food from their shelves. Although there are differences of degree, consumer surveys continue to show resistance to GM food.

The crop trials

In Europe, there have been no commercialisation approvals for GM crops since 1998, and only 18 consents – mainly for vaccines – have ever been granted. Although the legal mechanisms are there, a number of Member States have operated an informal – or unlawful – moratorium. They wished to await the negotiation of a new Directive which would allow them to take into account a broader range of consideration, including the direct and indirect effects of the crop, and its management, on the environment and biodiversity. That power came with a renegotiated Directive that was adopted in 2001. Some Member States wished also to await the introduction of a labelling and traceability regime, which seems likely to be approved later this year (2003). Some Member States apparently also wish to await the outcome of negotiations around the introduction of a legal liability regime, under which biotechnology companies and/or seed producers might be held strictly liable to compensate for environmental or economic damage arising from commercialisation.

It was against this background that the farm-scale evaluations of herbicide-tolerant crops were set up in the UK. They came about as a result of an agreement between the Government and the industry in order to assess the impact on farmland environment and biodiversity of the herbicide management regimes associated with the crops. This is done by comparing the effects of matching fields of GM and non-GM crops. Four GMHT crops are involved: winter and spring varieties of oilseed rape, beet (fodder and sugar) and forage maize. Between 65 and 70 fields, varying in size from 4 to 30 hectares have been planted with each of the four crops, and each field is split in two, one half being sown with an equivalent non-GM variety. Although this is an important project of ecological research, it is addressing a relatively narrow range of questions: whether the herbicide management regime associated with these crops, as compared to that used on the non-GM equivalents, has any effect on some aspects of farmland biodiversity (i.e., on the number and diversity of plants and animals).

Politically, however, the trials carry much more significance; indeed, they have come to assume an iconic political importance that greatly exceeds their carrying capacity. There are two reasons for this: first, that their political effect was to buy further time for the Government before it would have to make decisions whether to clear these crops for commercialisation; second, they introduced a spatial dimension to the use of the technology. GM crops were now being grown in people's villages and local environments.

Science, protest and popular culture became intertwined when both Lord Melchett and Tommy from the Archers deliberately damaged trial crops and were charged with criminal offences. Both were acquitted, Lord Melchett in real life and with greater dramatic effect even than Tommy Archer, only after a second trial when the first jury failed to agree. GM has assumed huge political significance, particularly in Wales (where the National Assembly's policy is as strongly against GM as it lawfully can be) and Scotland, where protestors spent last year in an encampment alongside a GM site at Munlochy on the Black Isle.

The trials are due to come to an end this year, and the first set of results is promised in the autumn. So the stay of political execution is running out. Decisions around commercialisation are looming. The UK Government has already this year received two applications for consent to import GM maize and oil seed rape, and various applications for cultivation consents have been lodged with the Governments of Spain, Belgium and Germany.

Regulatory models

In theory, there are several approaches which a government might take when designing a regulatory system for the introduction of new technology, within a range of possibilities from absolute prohibition to complete deregulation. I should like to examine three models.

Model One focuses on the product rather than the process. In the case of GM food, for example, a common and simple test might be that of substantial equivalence, which considers whether a new food product is substantially equivalent to analogous food products if any such exist. Hence it embodies the idea that existing organisms used as foods, or as a source of food, can be used as the basis for comparison when assessing the safety for human consumption of a food or food component that has been modified or is new. Under the current US approach, if there is no obvious difference between the assessed product and its natural counterpart, regarding its appearance, taste or selected chemical and nutritional properties, it is assumed to be equivalent and then no thorough testing is necessary.[4] The 'substantial equivalence' test was seriously criticised by the Royal Society of Canada, following an interdisciplinary study in 2001, as scientifically unjustifiable when used to exempt new products from full scientific scrutiny.[5] Nonetheless, the general principle resonates with GM: if two given products are identical, is it right that the production of one should be regulated, but not that of the other?

Model Two extends regulation also to the production process. This approach is common, for example, with industrial processes, where societies tend to regulate both the product (e.g., is it safe?) and the process (health and safety, human health and the environment). Again, there is a significant difference in approach between the USA and Europe. In Europe, releases of genetically modified organisms (GMOs) have been separately regulated since 1992 following the adoption of a Directive that year, which was revised and updated in 2001. Consent is required under Part B for the contained use of GMOs, and under Part C for their deliberate release into the environment.

Model Three adds a test of social acceptability: does this product, produced through this process, command broad support? Is it better than what we have already? Does it somehow improve the lot of society or some significant portion of it? This question is rarely posed in relation to consumer products. Social acceptability in a consumer society is a matter wholly for the market. It tends to be asked only when there are also appreciable social costs or perceived risks falling outside the regulatory structures outlined above that require a different balance to be struck. The regulatory paradigm of capitalist societies for new technology is that the manufacturer is entitled to bring a new product to market

---

[4] For Europe the test is more rigorous, under the EU Novel Foods Regulation 258/97 which provides for a statutory, pre-market safety assessment of all novel foods, including those obtained from GM sources.

[5] Royal Society of Canada, *Elements of Precaution: Recommendations for the Regulation of Food Technology in Canada* (January 2001).

provided there is no evidence that harm may be caused to human beings or the environment in its production or use. That is underpinned in world trade terms by the WTO. In many cases, the answer to the question seems to be self-evident: new technologies create added value and economic well-being. Bill Gates was never required to prove that the PC was a good thing. In land-use planning decisions, however, societies do ask the question: is this superstore, or this nuclear generating plant, actually needed? In relation to GM crops, too, the question might be, what contribution might this technology make to developing a more sustainable agriculture?

The current European regulatory paradigm is Model Two. The contained use of GMOs is tightly regulated; so too the deliberate release of GMOs into the environment. Other significant agronomic changes – such as the shift to autumn sowings of over-wintering crops – are not regulated at all, indeed may be directly subsidised, despite their adverse impact on farm-dependent bird-life. Although the new Directive pays some lip-service to broader ethical and social issues, the risk assessment it requires is primarily science-based. Herein lies a serious and continuing dilemma for the Government, because this approach risks disguising important social judgements behind narrow scientific frameworks.

Case-by-case regulation on a pan-European basis is not an appropriate forum for addressing the broader issues which have been raised by those who are concerned about the technology. Some of those concerns are about scientific issues: for example, gene flow from GM crops into wild relatives, how far this matters, particularly when it leads to gene stacking of more than one modified trait, and how risks can best be managed. Other concerns are about the broader ecological impacts of a GM agriculture, of weed-free and biodiversity-unfriendly fields. Other concerns are about the future character of British agriculture, and about the relationship between GM and organic farming. Studies of public attitudes indicate that there is a lack of public confidence in traditional narrowly framed, expert-based, quantitative approaches to risk assessment in fields such as GM crops and foods: hence the Government's reluctance, in light of recent experiences with BSE and foot-and-mouth disease, to ignore these issues.

The political elephant traps

This is exceptionally dangerous territory for politicians, not least because there is wide diversity of views amongst their leading representatives. It is difficult not to sympathise with the Government's current dilemma. It already has a regulatory regime which provides for rigorous case-by-case assessment, and it has surrounded itself with a plethora of expert committees and advice, including notably the Advisory Committee on Releases to the Environment, conveniently abbreviated as ACRE.

The habits of governments and politicians, when faced with such intractable problems, have changed with time. Up until the arrival of Mrs Thatcher in 1979, the usual reaction was to set up a Royal Commission. Their whole purpose was, as Harold Wilson once observed, to hold meetings, to keep minutes and to take years. By the time they reported, things would either have gone off the boil or there would be a new Government to sort out the mess.

Today, the response is more likely to be a clamour for a public debate. Rarely, however, is this taken beyond the bounds of political rhetoric. A few opinionated newspaper columns, a write-in campaign, a Select Committee grilling and some opinion polls to indicate when the clamour has died down and it is safe once again to make policy. Rarely has anybody thought through what a real public debate actually might be, what it is for, what it should involve, and how you know when you have had one.

The AEBC

On GM crops, we have had, or are having, a combination of the two approaches. There is a Commission, though it is certainly not Royal. It is the Agriculture and Environment Biotechnology Commission, set up in early 2000 following a Cabinet Office Review of the advisory and regulatory framework for biotechnology. Following an extensive consultation exercise, the Government decided to establish two new strategic advisory committees to work alongside the new Foods Standards Agency. They were to be the Human Genetics Commission (HGC), 'to advise on genetic technologies and their impact on humans'; and the Agriculture and Environment Biotechnology Commission, whose role was, rather ominously, 'to advise on all other aspects of biotechnology except food'.[6] The Government emphasised that its first concern in regulating biotechnology was to ensure the protection of human health and the environment, but followed that up with two qualifications: that the regulatory and advisory systems

---

[6] Cabinet Office and OST, *The Advisory and Regulatory Framework for Biotechnology: Report from the Government's Review* (May 1999).

needed to allow ethical and social concerns to be properly debated; and that they 'should not place unnecessary burdens on the industry or barriers to its development'.[7]

The composition of the AEBC was intended to reflect all of these perspectives, plus some more. Indeed, Ministers insisted on appointing no fewer than 20 members. They include leading members of environmental groups, molecular biologists, farmers (organic, conventional), academics, ecologists, bio-ethicists, representatives of consumer interests, people with experience of the biotechnology and seed production industries, and lawyers. I was not involved in any of the appointments: indeed, my own appointment to the chair came some time after all the other members had been appointed, Ministers having apparently already rejected a long list of candidates with qualifications considerably more impressive than my own.

It goes without saying that all the members are opinionated, articulate and passionate. It is also true that there seemed at the outset to be little common ground between them, beyond a genuine aspiration to make this unusual Commission work well. Early meetings were awkward and difficult, until we found our *métier*, and a proper appreciation of what we could realistically do. The Commission established a challenging commitment to open working. All our meetings are now completely open to the public. Much of the development work on major topics is undertaken by sub-groups of members (unlike the HGC, we do not co-opt non-members onto them), and draft reports are published on the Web as they develop. So is all my correspondence with Ministers. It is a vital step in re-establishing public confidence. The Commission has succeeded in gaining unanimity for all its reports so far, and in not losing any of its 20 members in its first three years, which must be almost unique for a Whitehall commission.

Crops on trial

The first major area of review by the AEBC was the crop trials.[8] This was a difficult study and a hard-fought report. The Commission consulted widely, held public evidence-taking sessions and large-scale public meetings. The Report's recommendations were clear, and were quickly accepted, almost in full, by the Government: although there had been several problems in the handling of

---

[7] Ibid., para. 7.
[8] AEBC, *Crops on Trial* (2001): www2.aebc.gov.uk/aebc/pdf/crops.pdf.

the trials, the Commission was clear that they should continue, but that their results when published should not simply be accepted by the Government as the last piece of the jigsaw in the run up to commercialisation of GM crops. The indicators of biodiversity being measured necessarily related only to the land in question, and over a relatively short time, and will require careful interpretation. The Commission had already commenced a study of whether changes were required in legal liability rules to secure the co-existence of GM and non-GM (including organic) crops; and it urged the Government to set up an independent review of further ecological data to complement the trials results, and to commit to an open and inclusive process of decision-making around commercialisation. It also urged the Government to assess GM crops not as a freestanding addition to existing agriculture methods, but in the context of the damage that so-called 'conventional agriculture' already does to the environment. Some 40% of the world's agricultural land, around 2 billion ha, is now seriously degraded, and even in the UK, where farmers are increasingly conservation-minded, intensification has damaged biodiversity in farmed landscapes. Our present agricultural practices are both environmentally and economically unsustainable.

The public debate

It was the AEBC's seemingly innocent recommendation for more inclusive decision-making which gave rise to the current activity of the GM debate. Like other authoritative advice to the Government in recent years, including the Royal Commission on Environmental Pollution's 1999 report on *Setting Environmental Standards*, the AEBC's report complained of the limitations to conventional methods of understanding public opinion in scientific and technological debates affecting the environment, and it advocated harnessing new deliberative techniques to this end. This is becoming a familiar theme: both the Royal Society and the Economic and Social Research Council are currently promoting research in this area.

A further element was highlighted by recent social research undertaken for the European Commission[9] specifically on public attitudes towards agricultural biotechnology, which indicated that public concerns could not be explained by

---

[9] *Public Perceptions of Agricultural Biotechnologies in Europe.* Final Report of the PABE research project funded by the Commission of European Communities Contract number: FAIR CT98-3844 (DG12 – SSMI) December 2001.

ignorance, or by a simplistic desire for zero-risk. Participants in a five-country survey, which showed a remarkable similarity of opinion between them, did not generally express entrenched opinions for or against GM: their responses were more nuanced and sophisticated. Ambivalence was the overwhelming feeling, since people recognised both positive and negative dimensions of developments in agricultural technology. However, an overriding theme was that of lack of trust in Government and regulatory institutions, in which the BSE experience was seen not as an exception, but as an example of the behaviour of such institutions.

Participants felt that policy-makers in their respective countries had not learnt from these experiences, in that they had not addressed any of the problems that had been demonstrated by the BSE affair. They therefore expected that these institutions would simply continue to behave in the same way with respect to GMOs, and other issues. They felt strongly that inherent and unavoidable uncertainties should be acknowledged by expert institutions, and taken into account in decision-making.

The Government understands that it has a difficulty with public trust, and it therefore committed itself to taking public opinion into account as far as possible in an open decision-making process, but set the committee the challenge of designing a programme. In a further report,[10] the committee proposed some fundamental ground-rules: the process should be run by a wholly independent steering board; the issues for the debate should be framed not by the Government nor the steering board, but by the people themselves; it should not be a mini-referendum but have both qualitative and quantitative elements, incorporating deliberative processes; it should be well informed. In short, the aim was to stimulate constructive public discussion of the social and ethical implications of the technology, developing a model which might be used for other areas of science-based decision-making.

In July last year I was invited to lead the process, and given a free hand to appoint an independent steering board. Designing the public debate so far has been a fraught process. Being truly independent of Government is an exposed position, especially when there are strong and public political differences on the issue, when one is spending public money, when all our meetings are as far

---

[10] AEBC, *A Debate about the Possible Commercialisation of GM Crops* (2002): www2.aebc.gov.uk/aebc/reports/public_attitudes_advice.shtm.

as possible in public, and when all our correspondence with Ministers is in the public domain. The Government has agreed to fund our activities to the extent of £500 000, a doubling of the original budget, and to extend the time-scale, not only to allow for crop trial results to be taken into account (though the timing of their publication has since been delayed) but to take account of forthcoming elections in Scotland and Wales.

More importantly, the Government has committed itself to respond publicly to the report of the debate, and to demonstrate also how its findings affect policy-making. This is an unusual step, but critical to the credibility needed to provoke people to participate in the debate.

What does a public debate involve?

The initial stages are now complete and are published: they include a specialist desk-review of work on public attitudes towards agricultural biotechnology around the world,[11] and a parallel study of international experience with different models of public engagement in policy-making around science and technology. We have drawn on them in developing our own methodology. Secondly, a series of 'issues framing' workshops have been held, true to our promise that the public, not us, should frame the issues for the debate.

This proved a fascinating experiment. We commissioned a specialist firm to set up 9 workshops, regionally distributed and with around 18 participants in each, selected on the basis of agreed sampling criteria. One workshop was specifically for people with some experience of the debate; the others were for people coming to it fresh. Indeed, they were not told the purpose of the meeting, beyond food and farming, but were invited to offer a list of what issues were uppermost in their minds. Not surprisingly, these included such diverse issues as the fire service strike, Saddam Hussein, house prices and the possibility of Manchester United losing. At the time of conducting the research, it was clear that the issue of GM was not salient in terms of dominating headlines, but the researchers found that people had views about it.

When asked to jot down words they would associate with GM, the responses were mixed and, at first sight, random. People frame issues in terms of their own lived experience. For example, most people nowadays are removed from the

---

[11] John Kelly and COI Communications Strategic Consultancy, *Desk Research for GM Debate Steering Board* (COI Communications, 2002).

processes of food production, so their attitudes to food quality, safety, etc., are founded in their relationships to supermarkets and shopping much more than they are by any relationship to producers and processors. The researchers were, however, able to break the attitudes into six categories: passivity, progress, suspicion, anxieties, possibilities and '*que sera sera*'. The issues they framed in terms of food, choice, information needs, uncertainty, targets, ethics and progress (which included such heads as science, technology, medicine, economics and ideology).

The researchers reported that the public do not believe that polarisation is to be avoided – indeed they want to hear all sides of the argument. They do, however, wish this process to go beyond adversarial posturing. They were hungry for the facts, though recognising that facts may not be readily agreed. Moreover, the public make it very clear that facts alone are not enough. They wanted also, if they were truly to engage in the process, to explore the reasons why GM was necessary, potentially useful or to be avoided.

The research also gave insights on the best approach to the design of a public debate and these have been built into the development programme. A resource toolkit is in preparation with the assistance of the Science Museum, and includes the development of factual information for the stimulus material, to include a video, a CD-ROM, printed materials and an interactive website. The overall programme is intended to have two main strands of deliberative processes: a so-called, 'narrow and deep' strand, in which people's engagement will be through facilitated focus groups, meeting on at least two occasions to allow thinking to mature and to be informed by reflection. Second, a three-tier approach to broader public engagement, with a series of large regional meetings, county-level meetings and network partnerships.

The science strand

Alongside the public debate there are two special strands of activity: a review of the science of GM, led by Sir David King, the Chief Science Adviser; and a review of the economics of GM, undertaken by the Prime Minister's Strategy Unit. Both of these activities are being undertaken in the public domain, and in a way which interleaves with the public debate. The steering board advised in the design and methodology of the two strands, and it receives regular reports from them.

It is critical that the three strands should interweave. The science review is in the hands of a specialist panel which also has social scientists amongst its membership. It has been holding public meetings, and itself meets in public. It has taken on board the public's framing of the issues. The panel's purpose is not to provide another inward-looking scientific review, but to investigate and engage with issues of public concern around GM crops and food. The principal issues have now been identified as food safety, gene flow and environmental impact. There are also two cross-cutting themes: horizon-scanning, and regulatory processes. Biology can be a fuzzy science: there are difficult issues of uncertainty and ignorance, which the Panel is addressing, and it is identifying areas where further research is needed.

The Strategy Unit have also developed an open, consultative approach in the development of its data sources and its methodology. They are considering a full range of possible scenarios for the future development of GM crops in the UK, including a 'no GM' scenario. This extends not only to the crops currently in trial but also to those in development that could be available within a ten-year time-frame. The categories of costs and benefits that are being examined include: those relating to the GM products industry; to the conventional (non-GM) and organic product industries; environmental costs and benefits – to the wider biotechnology sector and science base, and to the wider rural economy; and the impact on the ability of developing countries to make an informed choice about GM.

Nobody should underestimate the potential significance of the programme that is now under way, nor the openness nor interactive character of the process, nor the extent to which public attitudes and opinions are incorporated into it. The three strands between them embrace the wide range of issues that are of public concern and attempt to address them through an intellectual and participative framework that has been designed in consultation with the major stakeholders. People and organisations are starting to get fully involved in the process. It is a high-risk experiment for both the Government and the Steering Board, and like all experiments it may fail; indeed, it has been a roller-coaster ride over the past year. The alternative, however, a continuation of polarised exchanges around a controversial programme of decision-making, is even more difficult to contemplate.

Conclusions

So what conclusions should we draw from the current state of play? Back to C. P. Snow, who later wrote in reply to his critics:

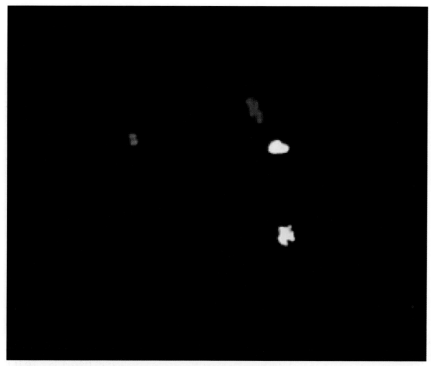

Panel a

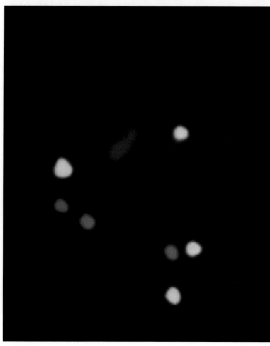

CHAPTER 6 PLATES Ia
and b. Visualisation of
selected chromosomes in
individual embryonic cells.
(Panel a) Normal distri-
bution in a male cell: two
copies of chromosome 1
(white) and one each of Y
(red) and X (blue). (Panel b)
Bizarre multiple chro-
mosomes in single cell
(same staining as in
Panel a) (© Imperial
College London).

Panel b

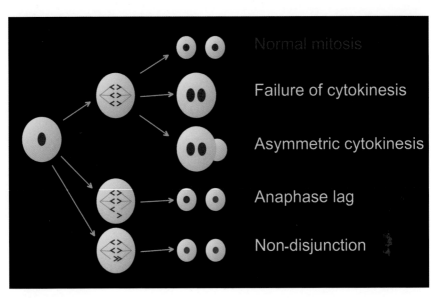

CHAPTER 6 PLATE II Various types of failure at mitosis. Red nuclei represent aneuploid nuclei (courtesy Dr Kate Hardy; © Imperial College London).

CHAPTER 6 PLATE III Australian cave painting.

Let me try again to make myself clear. It is dangerous to have two cultures which can't or don't communicate. In a time when science is determining much of our destiny, that is, whether we live or die, it is dangerous in the most practical terms. Scientists can give bad advice and decision-makers can't know whether it is good or bad. On the other hand, scientists in a divided culture provide a knowledge of some potentialities which is theirs alone. All this makes the political process more complex and in many ways more dangerous, than we should be prepared to tolerate for long.[12]

I would go further. Credible policy-making in science and technology today is more complex and more dangerous. It is not a matter of bridging between the divergent intellectual elites that represent Snow's 'two cultures', nor of educating the public in science, but of putting the necessary effort into developing more open political processes that will enable people to participate in intelligent and strategic deliberative ways to society's thinking about difficult issues, and allow Governments to overcome the otherwise inevitable lapse into formalistic, polarised and ultimately meaningless exchanges.

FURTHER READING

C. P. Snow, *The Two Cultures*, Cambridge: Cambridge University Press, 1998.

S. Martin and J. Tait, 'Attitudes of selected public groups in the UK to biotechnology'. In J. Durant (ed.), *Biotechnology in Public: a Review of Recent Research*, London: Science Museum, 1992, pp. 28–41.

Royal Society of Canada, *Elements of Precaution: Recommendations for the Regulation of Food Technology in Canada*, January 2001.

AEBC, *Crops on Trial*, 2001: www2.aebc.gov.uk/aebc/pdf/crops.pdf.

AEBC, *A Debate about the Possible Commercialisation of GM Crops*, 2002: www2.aebc.gov.uk/aebc/reports/public_attitudes_advice.shtm.

---

[12] Snow, *The Two Cultures*, p. 98.

6 DNA and reproductive medicine

ROBERT M. L. WINSTON

Institute of Developmental and Reproductive Biology, Imperial College, London

This essay will cover a number of topics in reproductive medicine, which are significant in the light of our recent understanding of DNA. I will discuss some aspects of the human embryo; the process of ageing in women; some aspects of gamete development and the failure of implantation; in-vitro-fertilisation (IVF) procedures, particularly pre-implantation diagnosis; cloning and transgenic animals; some aspects of mutational analysis in reproduction; and, finally, embryonic stem cells.

Prologue

I came across an amazing letter recently in the British Library. It is signed by Charles Darwin and I think has just been published by Richard Dawkins, FRS, in his latest book. It is addressed to Alfred Wallace, the person who coined the notion of survival of the fittest. In the letter he says:

> My dear Wallace, after I had dispatched my last note, the simple explanation which you give had occurred to me, and seemed satisfactory. I do not think you understand what I mean by the non-blending of certain varieties. It does not refer to fertility; an instance will explain; I cropped the Painted Lady and Purple sweet peas, which are very differently coloured varieties, and got, even out of the same pod, both varieties perfect but with none intermediate.

I think Charles Darwin did something quite remarkable here. What he is asserting in this paragraph is the particulate nature of inheritance (and possibly this predates publication by Mendel). It is really interesting that Darwin understood this; he obviously was a singularly great man and it is rather remarkable that this letter has only recently surfaced.

The human embryo

To step forward from Darwin's time over 150 years to discuss contemporary issues, which are of key importance to me as a reproductive physician and biologist – indeed, the enigma of the fragility of the human embryo and the fact that the human is one of the least fertile of all mammals, has driven so much of my own work over the last twenty years. Figure 6.1 shows a human embryo at the 8-cell stage, balanced on the tip of a domestic pin. This is just about 60 hours after fertilisation, just before compaction. It is some 100 microns across. The embryo has been gold-coated so it could be photographed using a scanning electron microscope. It appears to be completely normal, with perfectly formed, equal-sized blastomeres. Each of these blastomeres, the cells which make up the preimplantation embryo, is totipotential. Theoretically at least, each of these cells, if disaggregated from the others, has some potential for becoming a complete human being. Each of the cells has identical DNA. Yet, what is extraordinary is that we cannot tell by looking at the embryo – no matter how closely, or under what power of microscope – whether it is normal, or whether, indeed, it could develop into a child if transferred to the ideal environment. Looking at this embryo is about as instructive as looking at a person in the street and deciding whether he or she is intelligent.

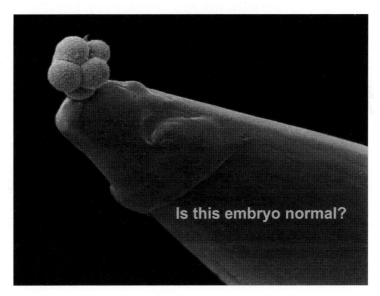

FIGURE 6.1   Scanning EM of an 8-cell embryo on head of a pin.

Patients who undergo reproductive technologies, IVF, are invariably told 'we've just transferred two wonderful grade one embryos'. In fact, such statements are meaningless and raise the hopes of the desperate, infertile patient – so much so, that, after embryo transfer, many women fantasise that they are pregnant. Over the last years, a great deal of my own research has been devoted to trying to understand how we might detect whether an embryo is normal and capable of becoming a human. Whether an embryo is normal is now an issue which I consider to be of key importance. It has become a matter of concern since IVF was started by Robert Edwards in Cambridge over 20 years ago.

Failure of implantation

One of the biggest problems is that most human embryos do not implant. Only 15–20% of human embryos implant after IVF and continue development. For a long time, it was not clear whether this was a problem with the embryo's environment, i.e. the lining of the uterus and the uterine cavity. The endometrium develops over about 16 days. During this period, there is just one brief moment, around about the middle of the menstrual cycle in a normal fertile woman, when suddenly on the surface of the lining of the uterus, specific mucus-filled structures called pinopodes appear. We think that implantation is generally possible only at the moment when these pinopodes are present in the uterus, when the conditions for cell adhesion are met. Certainly in the human, and to a great extent also in other animals, that synchrony between embryonic and uterine development is very important.

One critical factor is almost certainly the genes which are involved in the formation of the mucus which fills those pinopodes. Andrew Horne, a Ph.D. in our unit at Hammersmith, showed that, in a group of women who persistently experienced failed implantation, this mucus was likely to be abnormal. For example, after routine IVF with apparently normal-looking embryos, there was no detectable excretion of the pregnancy hormone, ßhCG, which is usually found from the first stages of attachment and implantation. In a number of these women, examination of their endometrium showed different expression of a particular mucus-like protein, MUC-1. This can be seen very clearly on immunocytochemical staining, when compared with the endometrium of fertile women undergoing IVF and other treatments. The gene responsible for the expression of MUC-1 (Figure 6.2) is very important in a number of tissues

FIGURE 6.2 Structure of the MUC-1 mucin gene. The region between exons 2 and 3 containing multiple tandem repeats of DNA is highlighted (© Imperial College London).

including the breast, as well as in the uterus. The interesting thing about this particular gene is that between exon 2 and exon 3 there is a region about 4–5 kb long – 4000–5000 letters of the DNA alphabet long – containing multiple tandem repeated sequences of DNA. Andrew found that when this region is shortened to 2.5 kb, as it is in a few women, then the protein appears to be abnormal and that, as far as we are aware, we have never seen an embryo implant – even though the endometrium develops completely normally and is indistinguishable from a normal endometrium on routine microscopy. We now know, incidentally, that MUC-1 is by no means the only gene which is associated with failed implantation in this way. Abnormalities of the Wilm's tumour gene (WT-1) have somewhat similar effects clinically. There are likely to be many others waiting to be discovered.

This kind of DNA technology is a revealing example of how examination of the individual's genome is going to become increasingly important when deciding on specific treatments. This kind of genetic diagnosis can be made rather cheaply during a preliminary investigation. Such a diagnosis might suggest that a patient requesting IVF should not have it on medical grounds, because the chances of implantation would appear to be zero, and one could avoid the cost to the health service. The finding also poses the question of whether, if an embryo is transferred at the same time as replacing the mutant protein in the uterus with the correct one, implantation might occur. However, that experiment has not yet been done.

From 100 human embryos transferred to the human uterus, only about 18 babies will be born at the end of 9 months (Figure 6.3). Most human embryos are lost in the first few weeks of life. Certainly the use of DNA technology has changed our understanding of this in a way that has been very important. What is clear is that the endometrial component that I have discussed causes only a tiny proportion of this failure. Humans produce zygotes with many defects.

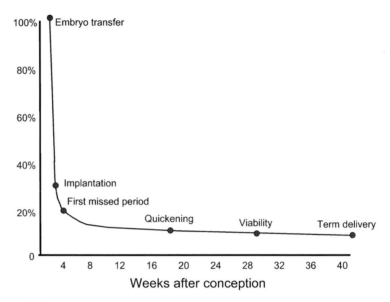

FIGURE 6.3   Success rate of embryo transfer and development, for 40 weeks after conception.

The ageing female

Some living creatures live much longer than others. The giant sequoia tree is one of the oldest living plants. Sequoia trees in California, which are over 3000 years old, are still fertile. Another example is the giant Galapagos tortoise. Harriet is such a female tortoise, living in a zoo 70 miles north of Brisbane, who was removed from the Galapagos Islands by Charles Darwin. This animal is 178 years old and won't mate so we don't know if she is still fertile. But many Galapagos tortoises of advanced age lay eggs.

Decreased fertility in the human female as age increases is of course well documented. Indeed, women are only capable of producing fertile eggs for about one third of their lives. Female age and reproductive ageing is now a significant problem in our society. More women reach absolute equality with men than ever before, in training academically and in skills. In medical schools now in Britain about 60% of students are female. In law schools the current estimate is that 52% of students are women. Women in the Western world are delaying childbirth until later in life (Figure 6.4), and this has important consequences. A little girl at birth has 2 000 000 eggs in her ovaries. By the

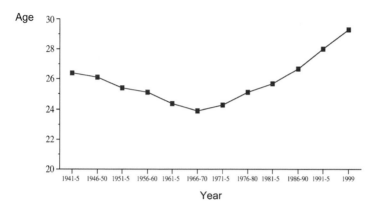

FIGURE 6.4 Maternal age at first birth, 1940 to date (source: National Statistics (www.statistics.gov.uk)).

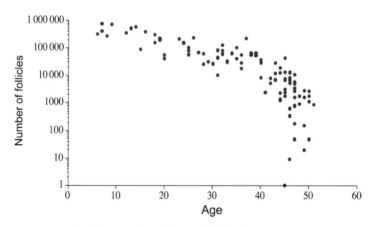

FIGURE 6.5   Declining numbers of oocytes/follicles from age 0 to 60.

time she reaches puberty she has about 300 000 left. Thereafter, she ovulates once a month, which would consume something like 400 eggs in an average female lifetime. Yet, by the time of the menopause 30–35 years later, she has virtually no oocytes left (Figure 6.5). Therefore, an additional process of oocyte depletion must be going on – this depletion is not simply due to ovulation.

This is a very serious issue for women – and an issue with which our society has not come to terms. It seems that women are hugely disadvantaged by their biology and they are hugely disadvantaged in many ways by the way the State handles these issues. It shocks me, for example, that no account of this process is taken when funding reproductive medical treatments. Most health authorities refuse IVF treatment to the women who most need it, women over the age of 36. We really should be considering how we might extend female reproductive life. There may be ways of doing this by manipulating the loss of oocytes in the ovary through the various genetic pathways which control it, or through changing the process of programmed cell death (apoptosis). However, at present, such treatments may still be dangerous, as can be seen from the following observations.

One of my American colleagues, Jack Cohen in New Jersey, has tried to get round this problem of oocyte death by refreshing the eggs of older women by transferring mitochondria from the cytoplasm of the egg of a younger woman under the age of 30 into the egg of an older woman over the age of 40. Mitochondria, which have 16 500 base pairs of DNA in them, are largely responsible for energy control and thus provide the energy which helps cell division. The theory is that if you placed fresh mitochondria into ageing eggs you might help drive the egg into cell division. Indeed, Jack Cohen managed to initiate embryo development, it seems, and this treatment did produce children. However, I think it was unfortunate that this was published as a success in a peer-reviewed journal in the way it was. It was only afterwards that we learned that some of these children had a deletion of the X chromosome, i.e. that their nuclear DNA was not always normal.

One very serious concern is that fertility treatment is an area of medicine where patients are often desperate to try anything which might help them. There are a number of therapies available – like vitamin treatments, drugs to improve endometrial development, modifications of IVF – where there is little or no evidence for their efficacy. At the same time, reproductive medicine is very often driven by frankly commercial interests. This commercialisation is a major problem for us, as it is with cloning. It risks bringing good science into disrepute, and we cannot afford that. Our society now has to be science-literate and we have to be extremely responsible about technologies produced from that science.

IVF procedures

To come to some more particular DNA problems in reproductive technology: one is that, in order to be successful at IVF, we drive the ovaries hard to produce more than one egg. One of the issues, which has not been properly focused on, is that in driving the ovary very hard with follicle stimulating hormone (FSH), nowadays usually genetically engineered, we run the risk of modifying the oocytes we are producing. Steve Franks and Kate Hardy in our laboratory have taken germinal vesicle stage oocytes from mice, after stimulating the mice with various levels of FSH. They have tried to mimic what happens in the human (in various IVF programmes, massive doses of these hormones are given). To make this experiment as pure as possible, prepubertal mice are treated and then the oocytes can be examined. They find that there is a rise in the number of mature eggs that reach metaphase 2. What is also very clear is that driving follicular development with high levels of FSH results in more eggs being abnormal, with fewer haploid eggs (which have one set of chromosomes as required just before fertilisation) and an increase in the number of eggs which have larger numbers of chromosomes or have missing chromosomes. This could be a fundamental problem in humans. We may get many oocytes after superovulation, but we may also get eggs that are 'less good'. This is likely to be only the tip of the iceberg. It would be very interesting to look at this phenomenon in the human, where I expect this effect to be much more profound. After all, mice multiply ovulate normally, whereas human females have a dominant follicle with a single mature egg. Therefore, producing multiple mature eggs is likely to be a good deal less physiological in large primates than it is in rodents.

What is interesting is that a very large number of human embryos are abnormal, but only a few of these actually develop further into foetuses and are ultimately born (Figure 6.6). Possibly about 20–25% of human eggs have abnormalities of their chromosomes and a further 20% of the very early embryos have chromosomal abnormalities as well. By the time the foetus is viable, most of these have been screened out by nature. For example, Down's Syndrome occurs in 1 in 800 births overall and Turner's Syndrome is rather rare as well (Figure 6.7; it is probable that a large number of miscarriages and embryos which do not implant have a deletion of the X chromosome). This is one of the very good arguments in favour of embryo research; it seems to me that until you get to the stage of successful implantation, the foetus cannot truly be regarded as a human. It has the potential in some cases of being a human. In fact, this 45% loss of embryos preimplantation is only part of the problem. There are many other reasons why embryos do not develop, even when they are chromosomally normal.

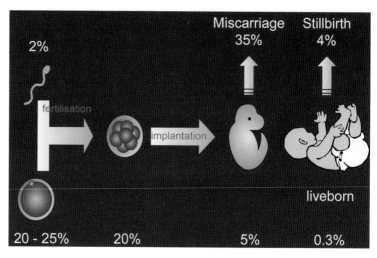

FIGURE 6.6   Incidence and time course of aneuploidy (courtesy Dr Kate Hardy).

Aneuploidy

0.3% newborns have a chromosomal abnormality

Surviving trisomies
Trisomy 21 (Down's Syndrome) - 1 in 800 births
Trisomy 13 - 1 in 20 000 pregnancies
Trisomy 18 - 1 in 10 000 pregnancies

Surviving monosomies
Turner's Syndrome - (monosomy X)

FIGURE 6.7 Percentage of newborns with chromosomal abnormality (© Imperial College London).

Preimplantation diagnosis

This brings me to the issue of preimplantation genetic diagnosis. Back in the late 1980s, Alan Handyside and I biopsied cells from the human embryo, taking a cell for DNA analysis. In those days, thermal cyclers to do automated DNA polymerase chain reactions (PCRs) were not available. I remember doing one early PCR experiment on DNA from single cells: it was very boring, using one water bath at 93°C and another at about 50°C, plunging the sample from one side to the other on a Sunday afternoon while listening to Shostakovich on headphones with a stopwatch for timing. Now, of course, this has been somewhat better automated, but these experiments led to the ability to carry out preimplantation diagnosis. We now can remove one or two cells from a human embryo, apparently with impunity. Even though the embryo has lost some of these totipotential cells, it makes up for them during growth. In fact, this kind of DNA diagnosis seems to be surprisingly safe. There have been no cases of abnormal babies being born amongst the several hundred babies that have now been born after application of this technique. So, we can analyse the following: we can determine the sex of the embryo, by staining for the X and Y chromosomes, frequently using chromosome 1, or others, as a control. We can diagnose sex with confidence, which is really how we started back in 1990; it led to the first babies being born. We can also check for the absence of some mutations leading to genetic diseases. One of our first successes was the birth of healthy twin girls whose elder brother aged 4 had died of a rather nasty X-linked metabolic disease, adrenoleukodystrophy. The girls are now 13 years old. At the time, and still rather sadly I think, the newspapers insisted on calling these 'designer babies'. They are no more 'designer' than I am. They simply have a single base pair difference from their brother who died. Apart from that, their DNA is a random mix from their mother and father. We did not design it and we did not choose it.

This method of chromosome analysis has become increasingly sophisticated. We are now able to look at cells in the embryo on the same day as we take the cell. Typically we get a diagnosis within 6 hours (this is an important point, because there are many reasons why we do not want to keep the embryos in culture). We can now stain most chromosomes in an embryo (Plate Ia). Work using a variety of techniques with multiple probes has now shown that only about one quarter of human embryos have completely normal chromosomes in every cell. Preimplantation diagnosis is a technique which not only gives rise to live babies, but also provides valuable insight into embryonic development. Very often we see a bizarre chaotic distribution of chromosomes in individual embryonic cells (Plate Ib). In the case shown in Plate Ib, there is one copy of the Y chromosome, four copies of chromosome 1, and three copies of the

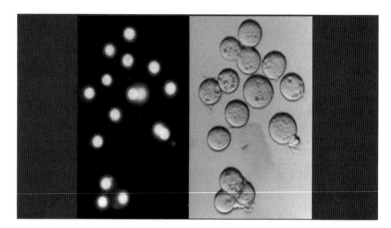

FIGURE 6.8 Multinucleate cells in 14-cell embryo with 16 nuclei (courtesy Dr Kate Hardy; for more details see: Hardy *et al., Journal of Reproduction and Fertility,* **98,** 1993, 549). 60% of day 4 embryos have binucleate, anucleate or multinucleate cells (© Imperial College London).

X chromosome. Almost certainly no single event could have caused this. It was probably a mixture of events, which included chromosome non-disjunction and problems during mitosis. Certainly, it is extraordinarily interesting that this phenomenon is quite common and human embryos are very frequently mosaic. Kate Hardy in the early days showed also that a number of embryos do not divide properly. For example, Figure 6.8 shows her picture of a 14-cell embryo that contained 16 nuclei. As human embryos develop more or less synchronously one would expect there to be 2, 4, 8 or 16 cells with a single nucleus each. In this case two of the largest cells have failed to divide, resulting in two cells with two nuclei each. Multinucleate cells in the human embryo are very common. We do not know why this occurs and if it is an environmental problem. The defects which occur may be failure of cell division, as in Figure 6.8, but you can also get non-disjunction where two chromosomes are segregated to one daughter cell and none to the other. Various other defects can also occur during cell division in the human embryo (Plate II).

Origins of aneuploidy

We have become increasingly interested in our laboratory in how chromosomal defects in the human embryo arise. They may occur at a very early stage. An interesting prediction was arrived at by a mathematical model, indicating that the defects could not arise because of the embryonic environment, but that they have occurred much earlier. Recent work has now demonstrated that this may well be the case and that many oocytes at a very early stage may have abnormal chromosomes.

Mosaicism probably occurs during early embryonic cell division. We expect that, if non-disjunction occurs at the 2-cell stage, then one of the two lineages will have abnormal copies of all these chromosomes thereafter. One of the things we have been able to do with embryos, quantitatively, is to observe how often it occurs in a particular embryo and thereby calculate whether this was at the first, second or third cell division. That is a very valuable process and gives rise to some very interesting ideas about what is going wrong with the human embryo. In our view, it is probably not just due to cell cycle check-points, but to other factors. One of the questions we are asking is whether or not the process of programmed cell death (apoptosis) controls this; whether, for example, the embryo is unable to express genes that cause these cells to die with the consequence that they continue to persist. Santiago Munné, publishing from the USA, showed that mosaicism occurs in embryos from women of all ages, but aneuploidy increases with advancing age. This observation would fit well the model that non-disjunction occurs later, which is really quite interesting.

I am turning now to another area of DNA, to the really important notion of trying to find the perfect embryo or the embryo that is most likely to implant. If we could transfer a single embryo to the uterus with some real prospect of a live birth at the end of it, this would be wonderful. It would make IVF much cheaper and much less risky. It would also, of course, eliminate the really big problem for health care services – the rise in the number of multiple births. Anthony Lighten, who was a Cambridge undergraduate and worked as a Ph.D. student in my laboratory, looked at the influence of the IGF group of genes. He was very interested because IGF1 is certainly something which the embryo is ready to receive for its growth and because the receptors are present in the human embryo. One question is: what would happen if you added this growth factor to the culture medium of human embryos? Would that actually help development? Sophie Spanos, another Ph.D. student, has used TUNEL labelling (terminal-deoxynucleotidyl transferase-mediated dUTP-biotin nick-end labelling) to look at apoptosis, or programmed cell death. In the absence of IGF, we found increased apoptosis in about 15% of the cells. Importantly, when we gave IGF1 growth factor, there was a reduction in the percentages of cells undergoing apoptosis. Now, at first glance that looks rather favourable because it would indicate that you are reducing fragmentation and therefore the number of abnormal cells. However, the real issue is: are we not rescuing cells in an embryo at the risk of that embryo? What we might well be doing is creating abnormalities, which persist in the cell lineages. This would seem potentially dangerous – it is troubling that so many groups around the world read papers on IGF and have added IGF to human embryo culture without any kind of control. This is something we will not do in embryos that we are intending to transfer into the uterus because we do not actually feel that there is enough information yet to do that with safety or impunity.

Another finding made by Anthony Lighten was to show that IGF2, another important growth factor, is imprinted in humans. This gene is expressed only on the paternal chromosome and is responsible for early growth. In a cloned animal like Dolly the sheep, however, IGF2 might not be expressed normally as the clone inherits both pairs of chromosomes from one parent.

Cloning of humans

That is why it seems to me unthinkable that we could be contemplating cloning of humans, such as Antinori and others have suggested; and the Raëlians, bless them. One would say 'bless them' if they were just flakey, but the trouble is that this is very dangerous for us. It is dangerous for every scientist. It is dangerous for members of the public who need to understand that the science is largely being pursued responsibly. It is a major problem because these things are unfortunately caught up in the press enthusiasm. They do us huge damage and it is something that we have not, I think, dealt with as scientists terribly well. We should be much more secure in saying that Antinori making monstrous pronouncements and as a bombast could be responsible for making monsters. Simply, what he is doing has not been scientifically validated. In any case, there is almost no possibility that his claims could be true. Interestingly, you may remember that on 5 April 2002 he claimed that he had an 8-week pregnancy of a cloned baby. Where is it? It is now (at the time of the lecture forming the basis of this essay) 28 February 2003, the fiftieth anniversary of the discovery of the DNA double helix. Antinori's 'human gestation' must be the longest pregnancy in human history. I think that it is quite scandalous that so much credibility has been given to that sort of nonsense. It does so much damage, because, of course, we as scientists are public servants. We are here to serve our society and hopefully most of us are here because we really believe in a small way we may try to better it.

Problems with IVF

Finally, we come to the issue of multiple embryo transfer, which represents an attempt to overcome the problem of low success rates with the single embryo transfer. This approach has led increasingly to the birth of triplets and multiple pregnancies (Figure 6.9). A typical intensive care cot costs £300–£1000 a day and the chances of triplets all being normal are not very high. Therefore, I think it is mandatory to try and find ways of growing better embryos safely but so far we have not been able to do this. The DNA technology that is now becoming available, and which is going to be important in this context, is the ability to look at a multiplicity of genes expressed in the embryo and thus to assess the quality of the embryo before implantation. That offers a very exciting possibility.

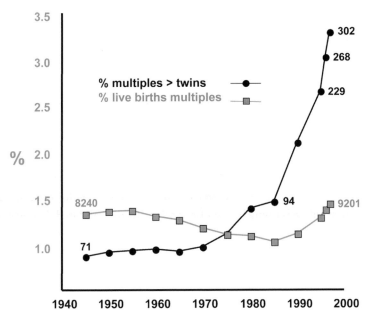

FIGURE 6.9 Multiple births in the UK.

Let me now discuss a couple of other issues before I come to a personal area of research that I am very involved with at the moment. One of the most serious issues that I think we have to take on the chin as reproductive physicians is the increasingly worrying reports in good journals, like the *New England Journal of Medicine*, of there being very high abnormality rates in some IVF units around the world. One particular publication from Western Australia, describing 527 births using routine IVF and 186 using intra-cytoplasmic sperm microinjection (ICSI), showed that the major defect rate was around 9.5%, where the background risk in the average population is around 4%. This, of course, is only one study, but the fact is that this odds ratio increased the risk from IVF more than two-fold in this practice. There is no question that these people are doing anything untoward. This is a very good unit. I think it is of concern to find out why some units are reporting figures this high and others are not. One of the things, incidentally, that happens in Western Australia is that a very large number of human embryos are frozen before transfer. They store them by cryopreservation. That may not be the only reason for the high incidence, because freezing on a global scale has not been a significant problem.

There is another paper published in the same journal by Dr Schieve's team in the USA. They looked at 42 000 human births after IVF and they compared them with 3.3 million naturally conceived babies during the same period in the USA. They have corrected for multiple births, prematurity, maternal age and, as far as possible, maternal health. What they found was that, in IVF units, the risk of having a small baby after IVF was 2.5 times that of the general population. That is very worrying. For small babies, it is not just a question of possible brain damage; there are many other issues there which need to be considered. This is something that we have known about in the UK for some time. The work of Sir David Barker and others has clearly shown that babies with low birth weights run much greater risks of having serious medical disorders when they reach middle age, probably because of the way they have to catch up with their metabolism. For example, if you are premature and born in Hertfordshire or parts of the North Country in Britain, from studies of births going back to 1913, you have a much higher risk of heart disease, atherosclerosis, stroke, diabetes, hypertension and possibly osteoporosis. Therefore it seems to me to be very important to be looking much more rigorously at the techniques which we are using in conventional IVF laboratories, to try and understand what happens to the human embryo in the rather artificial environment in which we place it.

I would like to mention one more example, from the work of Maria Tachataki, my current Ph.D. student. She has been looking at one particular gene, the tuberous sclerosis gene (TSC2). During the course of our experiments a few months ago, Maria ran out of human embryos. Human embryos are not easy to find. The scarcity of this material makes it very difficult and there is no substitute conventionally for the human embryo. Lymphocytes and other cells react differently. So she asked me to seek ethical approval to look at embryos we had in store; a number of patients have their embryos frozen and are happy to give them to research at a later date if they do not want them transferred to their own uterus. Ethical approval was granted and what we found was rather interesting. In one experiment measuring gene expression, there was a difference between cryopreserved embryos and controls in some embryos at day 2. This suggests that the expression of this gene TSC2, which is a tumour suppressor gene, may actually be reduced temporarily by certain freezing protocols which are widely used in general practice. Of course, it may just be a laboratory artefact, but I am prepared to say categorically that this kind of experimental work is essential if we are to understand more about the safety of these kinds of experimental procedure. Yet another reason why DNA technology is so important to us.

Embryonic Stem Cells

This brings me to plans for studying embryonic stem (ES) cells. So far, work on ES cells in our institution has only involved mice, though we also have clinical ethical applications pending for human cells. There is a wide interest in stem cells because of their potential to grow into any kind of tissue, so they can be transplanted into a patient with, for example, heart disease, Parkinson's disease or other problems. However, it seems that the embryonic stem cell is also a potential vehicle for answering key questions in reproductive medicine. They are embryonic cells and they can be replicated. They would be identical, and can be immortalised and would therefore have exactly the same genes. We could use them to validate cryopreservation and changes in culture conditions. We could also investigate aneuploidy, gene expression, and the ability to differentiate into different tissues in genetically identical embryonic cells as a way of validating IVF procedures. This would be an extraordinarily useful way of proceeding. The biopsy techniques discussed earlier might also be a very good way of trying to generate embryonic stem cells in an ethically justified fashion because, of course, the embryo can give up these cells with complete safety. It would be better to use embryonic cells than to use cells from the cord at the time of delivery, which probably are unlikely to provide anything like as good a resource (though this is arguable). In units like the Hammersmith, such cells could be used as a test bed for gene therapy applications. Patients are coming to us with genetic disorders and perhaps we could grow stem cells with a known genetic disorder to try to find ways of changing the genetics of that disorder.

Transgenic animals and transgenic spermatogonia

To come briefly to an area which has interested me sporadically. One big problem, which is essentially a reproductive one, is the inefficiency of making transgenic animals. It is really very difficult to make mice which carry reliably a particular transgene that you want to study. Of course, transgenic technology, first published by Gordon in the United States 22 years ago, has been one of the most significant advances in medical biology because it has led us to understand how genes work in a dynamic fashion and how they may change. It has enabled us to look at some mutations and a whole range of other things, but the problem has been that the technology requires the use of animal embryos which are difficult to obtain and difficult to manipulate.

Many years ago now, Carol Readhead and I hit on the strategy of trying to modify the testis instead of trying to change embryos, knowing that the male is producing sperm in great quantities. So, we have tried, both *in situ* and in spermatogonial stem cells taken from the testis, to modify them by injecting a DNA construct either with a viral vector, or with a range of other techniques through the *Vasa eferentia* in the testis. With the injection procedure, we can use either a dye marker or a marker consisting of little air bubbles. These then show up in the tubule system of the mouse testis, indicating consistent filling. We can either try to modify the spermatogonia after suppressing testicular function by radiation or by drugs, or we can take spermatogonia from immature animals, culture them outside the body and then try and modify them before replacing them into the testis.

One of the things we have tried to do (as has Brinster in the USA) is to transfer modified spermatogonia, the male germ cells, back into the mouse testis.

FIGURE 6.10   Transfer of modified spermatogonia back into mouse testis.

Figure 6.10 shows a father mouse with a dominant gene for a black coat, so all his offspring should have a black coat. Actually, however, three of his children do not have a black coat because he is producing sperm from spermatogonia that were replaced in his testis from another mouse. Spermatogonia transferred in this way will occasionally repopulate the testis. Unfortunately, the problem is that we cannot replicate this reliably. Most of the time, spermatogonia do not seem to link up with the Sertoli cells, and more concentrated work would be worthwhile in this area.

Young males, boys, who are undergoing cancer therapy, could have their spermatogonia stored whilst they are undergoing chemotherapy and radiotherapy. So it does have some immediate clinical applications. We have also transduced cells *in situ*, in the testis. In one group of experiments, we achieved a success rate of 86% in transgenic offspring, actually showing and expressing a reporter gene. Again, the problem is that it has been erratic and unreliable. We do not understand if this is a problem with the viral vector or whether it is a fundamental problem with the way the testis is being treated, so a lot more research needs to be done. At the moment, we have a licence to look at large-animal transgenics and we have been trying to do this with the pig. Why the pig? Well, if we could make pig transgenics safely with organs that would not be rejected by the human because of a genetic modification, we would have a wonderful resource for human transplantation. Pig heart, pig kidney, perhaps pig lung would all be about the right size for transplantation into humans. This is a possibility and in my view an alternative and much more satisfactory way than cloning of organs, because of the strange gene expression in cloned animals.

And, of course, if the animals are abnormal then there will probably also be strange gene expression in their organs.

Mutation analysis and possible gene therapy

One of the problems we face for the future is exemplified very well by a little boy, Manueli, who lives on the island of Sardinia. Manueli has beta-thalassaemia. This is a genetic disease due to a mutation present in about one sixth of the population of Sardinia. Manueli is a homozygote, so he has very severe anaemia, requiring repeated blood transfusions. The problem with these kids is that, even with the health care available in Sardinia, it saps about 60% of the island's health care budget. I know of course that, nowadays, attempts to transplant bone marrow into these children are much more successful, providing they are not iron-overloaded. However, the truth is that most of these children end up having blood transfusions that rescue them for a few weeks, but of course also slowly kill them because of the build up of iron in their tissues, in their liver, in their marrow and elsewhere. As a result, they develop all sorts of diseases, like diabetes.

Manueli has a characteristic bowed forehead. Many of his bones, because he is anaemic, make extra marrow to try and overcome the anaemia. Why not, if it were possible in a population where a particular gene sequence is so prevalent, modify the genes using one of the gene therapy techniques? Why not, if it were possible, inject into the testis of a little boy like this the right gene sequence to control the disorder? Well, this notion of gene therapy purely for human disease sounds good. Let us leave aside the issue of human enhancement, which a number of people, such as Jim Watson, have talked about, and let us just look at disease processes where there is a formidable ethical argument in favour. When you look at human skull bones from 3000 years ago that were dug up in Sardinia, it is very interesting to note that they show the same bowing. Therefore, this disease has persisted in this population over 3000 years and yet the population of Sardinia has survived in spite of it. Why? Well, partly, it is because these blood disorders are also protective. Thalassaemia protects against malaria. Supposing, for example, that the current epidemic of malaria got even worse with global warming and *Anopheles gambii* started to spread northwards into the Mediterranean where it was prevalent in the past, a whole population that did not have this genetic protection could be wiped out. So, the problem with this kind of DNA technology and reproduction is the very nature of its unpredictability. That is actually the kernel of this essay. The ability to analyse DNA in reproductive medicine ultimately argues for the possibility of manipulating the human and advancing our evolution. Clearly, if a species is defined by its DNA, then changing its DNA, which may become technically possible fairly soon, would at least in theory change the human species. It is very interesting to consider, if we believe that respect for human life is the cardinal ethical principle, what happens when we evolve ourselves into non-humans? Will humans still be protected and will human life still be sacred? Will we still be made 'in the image of God', and so on? These are interesting questions and the ones I think we need to be thinking about very seriously, in public lectures, in series like this, as well as elsewhere in public debate.

Epilogue

I was travelling through the Northern Territories of Australia recently when I came across a cave painting (Plate III), which an aboriginal guide showed me. He reckoned it was about 10 000 years old. The cave is in a very isolated part of Australia known as the Kimberley and the painting shows a crocodile-like figure, which is a spirit figure. It also, however, shows a prone figure, a woman, and underneath her is a prone child. I am willing to bet that this represents a death in childbirth, which must have been so common. What is so exquisitely moving about this painting is that these people felt anger and sadness and love and lust and ambition just as we do, and therefore to consider changing human attributes without an extremely good reason would seem to me unthinkable.

7 Genes and language

DOROTHY V. M. BISHOP

Department of Experimental Psychology, Oxford University

Humans and apes: close genetic relatives

Human DNA has a surprising amount in common with the DNA of other species. In his book, *Time, Love and Memory*, Jonathan Weiner described the amazement felt by researchers doing pioneering work on fruit-fly genetics, when they discovered the same molecular building blocks in fly and human brains. Recently, we have become familiar with the statistic that the human genome has over 98% in common with that of the chimpanzee. Such data provide impressive support for the theory of evolution, and vividly illustrate two key tenets. First, nature is adept at using old parts for new purposes. Lungs evolved, not by a sudden mutation in the genome leading to the growth of fully formed lungs, but by gradual modification of an organ that initially acted as a swim bladder in fish. Second, DNA is sometimes described as a blueprint for building a body, but this is misleading in implying that there is a one-to-one correspondence between genes and the body parts they build. A better analogy would be to think of the DNA code as more like a computer program, or, for those who prefer a more homely analogy, a knitting pattern. In both computer programs and knitting patterns, one has a list of instructions that lead to a final product. Both of these are codes where relatively small changes can lead to major differences in the final product. For instance, there may be an instruction that translates as 'do the following operation 10 times'. A change in the serial position of this instruction, or in the number of repeats that is specified, can lead to a very different result (see Figure 7.1). Just so with DNA. There are numerous genes whose function is to control the operation of other genes – switching them on or off, or determining how much protein they make,

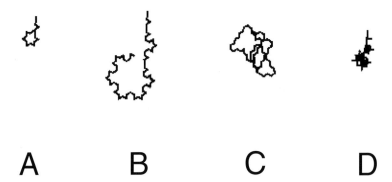

FIGURE 7.1 Results from a shape-drawing computer program. The programs used to generate figures A to D all contain 1177 characters of code, and are identical except for a single character.

or at what point in development they become operational. Minor changes to these regulatory genes can make enormous differences to the structure that is built. Once this is understood, one can see how it is that even small differences in the DNA of two species may lead to marked differences in their bodies.

Language: a major behavioural difference between humans and other animals

Despite the genetic similarities between man and ape, there is an enormous gulf in our intellectual abilities. In particular, language allows us to communicate complex ideas that go way beyond anything that can be achieved by other animal communication systems. We can talk about the past, the future, about hypothetical situations. We can entertain and inform one another with narratives. We can pass on knowledge to others so that a huge amount can be learned via the spoken or written word, without the need for direct experience. Language comes so naturally and easily to most of us that we barely consider it, but it is a quite astonishing achievement, and one that remains poorly understood.

It is important to appreciate just how great the gap is between humans and our cousins, the apes. Other species do communicate using a variety of means, including vocalisation. Dorothy Cheney and Robert Seyfarth have carried out elegant and painstaking studies of vervet monkeys, demonstrating that they

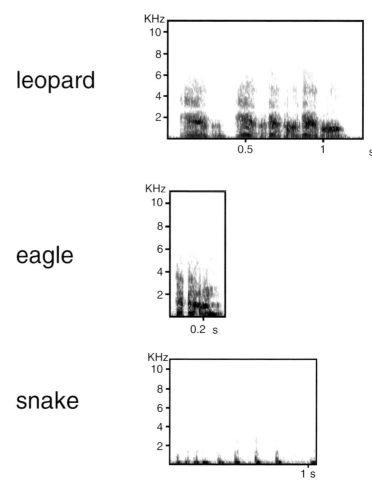

FIGURE 7.2 Alarm calls given by vervet monkeys in response to different predators (reprinted, with modification, from Figure 4.4 in Cheney and Seyfarth, *How Monkeys See the World*).

use distinctive vocalisations in response to different predators (see Figure 7.2). However, these calls differ from human language in two important respects. First, they are not truly symbolic. A symbol can refer to something in a range of contexts – for instance, the word 'fire' can be used to talk about a fire when none is present. Primate vocalisations appear much more context-bound – more akin to someone shouting 'fire!' in an emergency. Second, human language uses ordered sequences of words to express entire propositions, something that is not seen in primate vocalisations.

In the 1960s, there was optimism that apes may have the mental capacity for language acquisition, and just be limited by the configuration of their vocal tracts, which are not adapted for the production of articulate speech. Accordingly, a number of attempts were made to train apes to use nonverbal languages, such as American Sign Language (a gestural language developed in the deaf community) or artificial visual symbols. These studies revealed that chimpanzees could use symbols, but this was only achieved with laborious training, and there was little evidence of any ability to string words together in sentences to express more complex ideas. The superstar of the ape language studies is Kanzi, a bonobo (pygmy chimpanzee) who has been studied over many years by Sue Savage-Rumbaugh and her colleagues. Two things are especially impressive about Kanzi. First, he picked up knowledge of the meanings of visual symbols by observing another bonobo being trained, rather than by direct instruction. As an infant, he accompanied his adoptive mother as unsuccessful attempts were made to train her in the use of a symbol board. On an occasion when the mother was absent, Kanzi surprised the researchers by demonstrating skill in the use of the symbols. Second, Kanzi can understand a fair number of spoken words, and, even more impressively, can respond appropriately to spoken sentences with a level of accuracy comparable to a 2-year-old child. Some examples of his responses to commands are shown in Box 1. A great deal of heated debate has been generated on the question of whether Kanzi can be said to have mastered language. Savage-Rumbaugh argues that people are reluctant to credit his achievements because they are uncomfortable with the idea that apes have more in common with us than we thought. Nevertheless, a more valid reason is that, however impressive Kanzi may be, he does illustrate that one aspect of language, syntax, is beyond the grasp of even the most gifted ape.

What makes syntax so special?

The linguist Derek Bickerton argues that what Kanzi has mastered is protolanguage. He can use words to represent things and actions, and he can string them together in a meaningful fashion. Simple combinations of words can be interpreted by using general knowledge and contextual cues. In this regard, Kanzi is like a typical 2-year-old human. The 2-year-old child differs, however,

Box 1

Testing language comprehension in an ape

Savage-Rumbaugh and her colleagues assessed Kanzi's comprehension by giving him spoken commands and observing his responses. On each trial a group of six random objects were placed in front of Kanzi. Other objects were placed in strategic locations out of sight. Two assistants were in the room with Kanzi, but they wore headphones through which loud music was played, so that they could not hear the test sentences. Test sentences were spoken by an experimenter who sat behind a one-way mirror to prevent any unintentional nonverbal communication with Kanzi. Examples of Kanzi's responses to commands are given below. The same test was administered to a 2-year-old child, whose overall level of performance (66% correct) was slightly lower than that of Kanzi (72% correct). (For full details see E. S. Savage-Rumbaugh, J. Murphy, R. Sevcik, K. E. Brakke, S. Williams and D. Rumbaugh. 'Language comprehension in ape and child', *Monographs of the Society for Research in Child Development* **58**, 1993).

Examples of commands that were correctly responded to:

- 'can you put the ball on the pine needles?'
- 'make the snake bite the doggy'
- 'get the melon that's in the potty'

Examples of commands that did not elicit a fully accurate response:

- 'go to the sink and get a knife'
  Kanzi takes a melon from the array and runs to the sink with it
- 'give the dog picture to Kelly'
  Kanzi gives the keys to Kelly
- 'put the rock in the bowl'
  Kanzi puts the rock in the water.

in that, over the course of a couple of years, it develops the ability to convey complex ideas that can only be expressed unambiguously by using word order and/or special grammatical words and word-endings to denote the relationships between entities. Bickerton uses as an example the sentence: 'the boy you saw kissed the girl he liked'. If we were to try to express this idea in proto-language, it might sound something like 'boy you see kiss girl boy like'. This, however, would be ambiguous: Who kissed whom? Who likes whom? Did the kissing and liking happen in the past, present or future? Syntax gives humans the ability to combine meaningful units in ways that express, in a precise fashion, meanings that are more complex than just the sum of the component words.

A key characteristic of syntax is that it is recursive. A clause may consist of a subject (e.g. 'the boy'), a verb ('kissed') and an object ('the girl'). Both subject and object, however, can be expanded into a whole clause ('the boy you saw'; 'the girl he liked'), and these can in turn be expanded again (e.g. 'the boy you saw when you came to Oxford kissed the girl he liked instead of the one he arrived with'). Syntax allows us to recover the underlying hierarchical structure of such sentences to arrive at an unambiguous reading.

People have implicit knowledge of rules of syntax, though they have difficulty in stating them. We know, for instance, that it is acceptable to say 'the boy you saw when you came to Oxford kissed the girl he liked' but not 'the boy when you came to Oxford you saw kissed the girl he liked' or 'the boy you saw when you came to Oxford kissed he liked the girl'. It is important to stress that when linguists and psychologists refer to humans' underlying knowledge of syntax they are *not* talking about grammatical correctness of the kind that is taught in schools. A Cockney may, for instance, accept a sentence such as 'them boys what you saw when you came to Oxford kissed the girl'. Different dialects have different criteria about what is an acceptable English sentence: the important thing is that speakers of the same dialect will generally agree about what is and is not acceptable, indicating that their linguistic behaviour obeys a common set of rules.

So what happened in the course of evolution to give us our facility in language? Here there has been considerable division of opinion. The most influential linguist of the twentieth century, Noam Chomsky, proposed that humans have innate knowledge of certain grammatical principles. According to Chomsky we do not *learn* language any more than we learn to breathe or to pump blood around the body. Our language faculty is seen as a 'mental organ', analogous to other bodily organs, such as heart and lungs, which are innately specified to carry out a particular function. At first glance this seems frankly preposterous – children do not spring from the womb with full knowledge of language. Furthermore, a child growing up in Kyoto will speak

Japanese, whereas one growing up in Cambridge will speak English. The grammars of these languages are quite different: in English, sentences usually obey Subject-Verb-Object word order ('John hit Fred'), whereas in Japanese, Subject-Object-Verb word order is normal ('John-ga Fred-o but-ta'). Of course, Chomsky is aware of these undeniable facts. However, he argues that what we observe in the developing child is not language *learning*, but rather *triggering* of language knowledge by exposure to the ambient language. Although different languages have superficially different syntax, at an abstract level we can identify a core set of underlying grammatical principles that are common to all languages – Universal Grammar – and it is these that are an innate part of human makeup. Chomsky had rather little to say about how this innate ability evolved, but, insofar as he does comment on it, he appears to adopt what has been termed (rather disparagingly) the 'hopeful monster' account of evolution, namely that a single genetic mutation suddenly appeared that happened to make language possible.

Why does Chomsky advocate such an extreme position? I would argue that at the heart of his arguments are two fundamental misconceptions: a wrong assumption about *what* children learn, and a misunderstanding about the kinds of learning that are possible using general mechanisms that do not have prespecified knowledge of grammatical rules. First, Chomsky's view of *what* is learned is derived from the study of adult grammars. His famous sentence 'colourless green ideas sleep furiously' was used as evidence that grammar is independent of meaning – any competent speaker of English will agree that this sentence is grammatical though meaningless, whereas 'green sleep colourless ideas furiously' is neither grammatical nor meaningful. Chomsky assumed that because *in adults*, syntax can be dissociated from meaning, the task confronting the child was to learn grammatical rules, without any recourse to meaning. On this view, having identified words as belonging to particular classes, such as noun, verb or adjective (and Chomsky remains silent on the question of how this is achieved), the child must induce underlying rules that specify which sequences of these abstract grammatical categories are legal, and which are not. There is a branch of formal mathematical learning theory that is concerned with specifying the necessary and sufficient conditions for such learning to take place, and it is clear that it is only possible to deduce the grammar of a natural language if one has exposure to both grammatical and ungrammatical sentences, with instruction as to which is which. Clearly, this is not what happens in language acquisition, where the child has little exposure to ungrammatical input, and, when it does occur, it is not labelled as ungrammatical. The only possible conclusion, according to Chomsky, is that our knowledge of grammar is not learned, but innate. However, an alternative is that the argument is proceeding from the wrong premises. There is mounting evidence that young children remain blissfully unaware of syntactic structures during the first couple of years of life, learning their language word by word and phrase by phrase in a piecemeal fashion, and depending heavily on meaning and context to communicate. For instance, Michael Tomasello has shown that 2-year-olds do not behave as if they have a category of 'verb'. They may have various verbs in their vocabulary, each of which could potentially be used in Subject-Verb-Object constructions, but use, for instance, 'move' only in V-O frames ('move brush', 'move broom' or 'move tray'), 'draw' only in S-V frames ('Mummy draw', 'me draw', 'dolly draw') and 'wash' only in S-V-O frames ('Mummy wash hair', 'me wash dolly', 'David wash hands'). The impression is that there is item-specific learning, with each verb initially being used in a rather restricted set of sentence frames to express a particular kind of meaning. Once children have acquired some critical mass of linguistic knowledge, they can identify patterns in the input and extract the more abstract underlying syntactic structure. This, however, need not be an all-or-none matter; computer simulations of learning by neural nets shows that rule-like behaviour can emerge from simple networks that acquire probabilistic associations.

Does this mean, then, that language is just like any other skill – riding a bicycle, or playing the recorder? We would not, after all, assume that humans have evolved some special genetic endowment that allows them to acquire these skills. Some psychologists have argued that language is no different from our other abilities, and that Chomsky's notion of a specialised 'language organ' can be dismissed altogether. However, two problems confront anyone who takes that stance. First, we need to explain why apes have failed to make the leap into syntax, despite their relatively advanced cognitive skills. Second, although neural network simulations have forced linguists to rethink the nature of what children learn, computer models have to date been singularly unsuccessful in simulating the process of language acquisition except in very simple 'toy' situations, which use a small vocabulary and a limited set of syntactic constructions. Neural networks are good at pattern detection, but spoken language is different from other kinds of patterning, in that its underlying structure is hierarchical, yet its patterning occurs over time (i.e. words follow one another in linear sequence). The recursive nature of syntax means that linear order may be a misleading cue to relationships between words. Thus in a sentence such as 'the elephant pushing the boy is big' the word 'big' occurs in close proximity to 'boy', yet it modifies the noun 'elephant'. This is one aspect of language that humans find trivially easy to pick up, yet computers have real difficulty with. For computer simulations to have even modest success in abstracting the underlying hierarchical structure of language, the neural network has to be set up with a particular kind of structure – as a recurrent network – and to be exposed first to simple sentences before proceeding to more complex structures.

Specific language impairment

Language is difficult to study precisely because it is specific to humans. With many other human characteristics, we can gain insights into underlying biological mechanisms by animal experiments, but this is hardly feasible in the case of language. Instead, we must rely on nature's experiments – people whose language abilities are affected by congenital abnormalities, or acquired brain injury. My particular focus of interest has been in children whose development appears generally normal, except that they fail to master language in the same way as their peers. Such children are typically slow to get started in talking, but do not, like many 'late bloomers', catch up. Rather, they persist in using language that may seem immature or abnormal. For instance, a 4-year-old may talk in simplified 2-word utterances saying 'want ice-cream' rather than 'I'd like an ice-cream.' In these children, grammatical errors are not just dialectical variants: the child's language is regarded as abnormal by other members of the same speech community. Often, in the early years, the production of speech sounds is also immature or abnormal, and so a phrase 'goldilocks and the three bears' may come out as 'doedilot an dee fee baird'. Usually, the most obvious problems are in formulating utterances, but more detailed testing often reveals some difficulties in understanding as well as producing language. For instance, the child may have unusual difficulty working out who is doing what to whom, and which is big, in a sentence such as 'the elephant pushing the boy is big' (see Figure 7.3). Children who have disproportionate difficulty in mastering their native language are said to have Specific Language Impairment, or SLI.

FIGURE 7.3 Item: 'The elephant pushing the boy is big' from the Test for Reception of Grammar-2, Standardisation Edition. The testee is required to select from the array the picture that matches the spoken sentence (reproduced with permission by The Psychological Corporation Limited. The Psychological Corporation, 2002, 2003; all rights reserved).

SLI has been of especial interest to those engaged in debates about the origins of human language, because it appears to confirm the idea that language is special, and not just part of our general intellectual endowment. If a child can develop normally in all respects except for language acquisition, this would seem to indicate that language is not just part of our general intellectual ability, but does require some specialised skills – in which children with SLI are deficient.

Genetic influences on SLI: evidence from twin studies

For a long time, it was assumed that SLI was caused either by subtle brain damage, perhaps incurred around the time of birth, or by faulty parenting. Earlier this year there was extensive press coverage in the UK of claims that developmental language difficulties were increasing because parents no longer talked to their children. However, these conclusions seem based on opinion rather than scientific study. In the 1980s and 1990s study after study appeared showing that the main difference between children with SLI and those with normal language was that the former came from families where other relatives also had language difficulties. Not all relatives were affected: in the same family one might find two children with excellent language skills and two with SLI. Findings such as this suggested that there might be an inherited predisposition to language difficulties, and raised hopes that the study of SLI might help us uncover the origins of human language, by leading us to a gene or genes that give us our distinctive skills in communication.

Aggregation of a disorder in families suggests that genes are involved, but it is not conclusive, because family members share many environmental circumstances and experiences, as well as having similar genotypes. Twins provide a natural experiment that allows one to disentangle to some extent the effects of shared environment from shared genes. Twin studies have attracted enormous publicity because of reports of astounding similarities between identical twins who have been reared apart, yet are reunited in adulthood and found to have the same taste for chocolate chip cookies, the same irritating mannerisms, and the same preference for women called Dolores. However, most twin studies are not like that. Rather than focusing on the rare cases where twins are reared apart, we look at twins growing up together in a normal family environment. Because they have so many experiences in common, we expect twins to resemble one another. The crucial question is whether identical twins resemble each other *more* than fraternal twins. Identical twins are formed by the splitting of a single fertilised egg. They are genetically the same, so any differences between them must be due to non-genetic factors, including such influences as the prenatal environment in the womb, health and nutritional status after birth, and the language they hear from parents and teachers. Fraternal twins are no more similar genetically than regular brothers and sisters – it is often stated they have 50% of their genes in common. This may seem rather puzzling when one considers that we are supposed to have over 98% of our genes in common with a chimpanzee. However, the 50% figure mentioned in twin studies is really a shorthand. It means 50% of *those genes that show variation from one person to another*. A very large proportion of the human genome is the same for all human beings – these are genes that regulate important bodily functions, and everyone will have two identical copies (one from each parent). If they mutate, those functions do not work, and the organism does not survive. Because these genes are the same for everyone, they cannot explain differences between people, and so behaviour geneticists – the people who do twin studies – ignore them. However, some genes, such as those that determine blood group or eye colour, come in different versions (alleles) – these are known as polymorphic genes. If there are genetic variants that are associated with disorder, then we would expect to find that an identical twin of an affected child should be more likely to have the same disorder than a fraternal twin.

Several twin studies have been conducted on SLI. They all find that concordance for disorder (i.e. the proportion of twin pairs where both twins have disorder) is substantially higher for identical than fraternal twins, consistent with the idea that genes are implicated.

The KE family and discovery of the FOXP2 mutation

When a disorder runs in families, we can look for telltale patterns of inheritance that provide clues as to genetic mechanisms. Suppose there is a single gene which has undergone a mutation, and that a single mutated copy of the gene is sufficient to cause disorder; a parent who has the disorder has a 50% chance of passing the defective allele on to a child. Other patterns of inheritance are predicted in the case where two defective alleles are needed to cause disorder, or where the defective allele is carried on a sex chromosome.

In 1990 a family was described (the KE family) in which SLI was found in three generations, with a classic pattern of inheritance consistent with a single dominant allele that caused disorder (see Figure 7.4). The initial account focused on the speech difficulties of affected individuals. This corresponds to a 'verbal dyspraxia' – difficulty in producing accurate sequences of sounds so that speech is frequently mispronounced and hard to understand. Subsequently it was shown that affected individuals also had particular problems with syntactic aspects of language, leading to excitement that they might lack the innate facility for language that had been postulated by Chomsky. Needless to say, those who contest the existence of such an innate facility were quick to argue against such an interpretation, noting that the problems of affected individuals extend beyond syntax. The team at Great Ormond Street Hospital who have done the most in-depth analysis of speech and language functions in the KE family have argued that the difficulties with syntax may well stem from their limitations in speech production, rather than reflecting lack of specialised brain

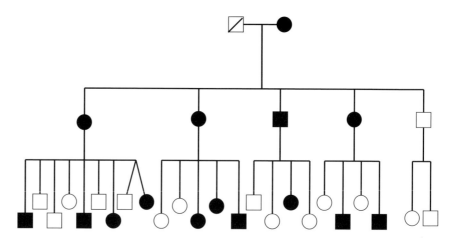

FIGURE 7.4 Pedigree of the KE family. Black symbols denote affected cases, and white symbols unaffected cases. Males are shown as squares, and females as circles. The first generation (an affected mother and unaffected father) are shown in the top row. The second row shows their 5 children, 4 of whom were affected. The third row shows the grandchildren, 10 out of 24 of whom were affected.

mechanisms for syntax. There is, however, a problem with this argument, and that is that the syntactic difficulties are evident in written as well as spoken language, and affect ability to comprehend syntactically complex utterances as well as to produce them. In this regard, members of the KE family differ from children who have purely physical difficulties in speech production, e.g. those with cerebral palsy, who have little difficulty in comprehending sentences such as that illustrated in Figure 7.3. The close relationship between syntactic and speech difficulties in affected individuals from the KE family could arise because adjoining brain regions are affected by the genetic defect, or it could indicate that both speech and syntax require similar kinds of analytic systems. For instance, it could be that we need a neural system organised like a recurrent network in order to extract hierarchical structure from a linear string of symbols – whether these are speech sounds or individual words.

The pedigree of the KE family strongly pointed to a single gene associated with SLI, so the next step was to try to identify the gene. It is easy to underestimate the difficulty of this task. With the completion of the Human Genome Project, many people assume we can just hunt through the genome until we find a gene that is different in affected people, but there are far too many genes for this to be feasible. A more efficient approach is to capitalise on the fact that DNA is full of noncoding regions – more colloquially known as 'junk DNA': sequences of base pairs that appear to serve no useful function (i.e. they are not genes and do not control the manufacture of proteins). Because these sections of DNA are nonfunctional, there is no selection pressure against mutations, and so there can be considerable diversity from one person to another in the DNA sequences that are found. Thus one can look at a pair of siblings, compare their DNA at these sites with that of their parents, and work out which parent contributed each section of DNA, as it is so distinctive. Junk DNA does not do anything, but it occurs close to regions of DNA that do form part of the genetic code. Regions of a chromosome that are close together tend to be inherited together, so if two people inherited a given section of junk from the same parent, the likelihood is high that they also inherited nearby genes from the same parent. Thus the junk DNA can be used as a kind of signpost or marker that allows one to identify whether two related people are likely to have inherited the same allele in a coding region of DNA. This is the logic of linkage analysis. One looks for marker regions that are shared by a pair of affected relatives with a greater probability than expected by chance. One can then do a more refined search of the allelic variation in genes in that region of a chromosome, to look for a specific allele that co-occurs with the disorder at above chance levels. This approach was used by Tony Monaco and his colleagues in Oxford to home in on a region of chromosome 7. Once linkage analysis had shown them where to concentrate the search, the Oxford team found a gene in that region that showed perfect association with disorder. That is to say, the DNA of this gene – FOXP2 – took one form in all the unaffected family members, but differed in a single base pair in all the affected members. Furthermore, the version of the gene seen in the unaffected members was the same as that seen in a large number of unrelated individuals from the general population. The Oxford group linked up with Svante Pääbo and his team of evolutionary geneticists in Leipzig to conduct a study comparing FOXP2 in chimpanzee and human. They found that the DNA of FOXP2 is identical in the two species for all but two base pairs. Furthermore, for the region of FOXP2 that differs between humans and chimpanzee, there is little or no polymorphic variation in normal humans. This indicates that a mutation in this gene occurred some time after the human lineage separated from that of the common ancestor with the chimpanzee, and then rapidly became fixed in the human population, indicating that it carried a substantial reproductive advantage.

The Oxford–Leipzig team have been at pains to point out that FOXP2 is not a 'gene for language'. It is a regulatory gene that can be found in a range of other species, including mice, and it affects the development of a range of bodily organs, including lung, heart and gut as well as brain. However, taken together, the evidence from the KE family and from the human–chimp comparisons supports the idea that the change in FOXP2 that occurred in human evolution during the last 200 000 years played a major role in the development of brain regions that are crucial for language. The DNA change is tiny, but, just as in the example of Figure 7.1, a small alteration to the genetic code can have potentially large consequences. If we can understand how FOXP2 functions to control early brain development (and work on this topic is under way) then this could throw light on the key brain differences that provide us with our language capabilities.

Beyond FOXP2: genetic influences on SLI in general

The KE family have proved remarkably informative, but they are not typical of SLI. Tony Monaco's team have screened many other families affected with SLI and to date found no other cases with the same mutation as the KE family. Clearly, though of great theoretical importance, FOXP2 is not a general explanation for children's language difficulties.

Furthermore, the clearcut pattern of inheritance seen in the KE family pedigree is the exception rather than the rule. Most children with SLI have a family history of disorder, but a typical finding is that the percentage of first-degree relatives (i.e. parents or siblings) who are affected is around 25%, rather than the 50% seen in the KE family. In addition, concordance for SLI in identical twins falls short of the 100% that you would expect if the disorder were solely due to a mutation similar to FOXP2. This means that environmental, as well as genetic, influences must be taken into account when looking for causes of SLI. It suggests that there may be genes that put the child at risk of SLI, but the manifestation of that risk depends on environmental experiences. These potentially could include a host of factors ranging from the prenatal conditions experienced *in utero*, through to the quality of language in the home, diet, health or schooling. There are many examples of such causal mechanisms in physical medicine; e.g. people with a genetic risk for developing emphysema may only develop serious disease if they smoke cigarettes. The existence of such interactions between genes and environment makes the task of the molecular geneticists harder because it means that, rather than looking for one-to-one relationships between genes and disorder, they have to work with probabilistic associations. In such situations, the task can be made more tractable if one can find some way of identifying people who carry the risk gene, even if they do not appear to have the disorder. In the field of language, good candidates will be measures that can reveal underlying deficits in people with a past history of language impairment, as well as in those who still have obvious signs of SLI. In my own work on SLI, I have adopted this approach, looking for language measures that may act as a marker of a person's risk status.

This line of research has led to some promising results. In one twin study, I focused on a simple task known as 'nonword repetition', in which the person is required to repeat back a nonsense word composed of an unfamiliar string of speech sounds, such as 'perplisteronk' or 'hampent'. This task poses challenges similar to those involved in learning new vocabulary: one has to decompose the acoustic signal into a string of familiar speech sounds, remember these in the correct order, and then articulate them. The psychologists Susan Gathercole and Alan Baddeley have suggested that humans may have evolved a specialised phonological memory system to do this task, as part of the language-learning faculty. We found that the task was not only very difficult for children with SLI, it was also often failed by children who had a history of slow language development but whose problems had resolved. In this regard, it seemed to be a promising behavioural marker of underlying genetic risk, and we went on to show that the twin data did indeed give striking evidence of a genetic component to this skill. Identical twins resembled one another strongly in their nonword repetition skills, whereas the similarity between fraternal twins was much less. We therefore suggested to our colleagues in molecular genetics that it would be worthwhile doing a linkage analysis using the nonword repetition test as an index of language impairment, rather than more traditional clinical criteria. This worked out nicely: in a study of families with two children affected by SLI a significant linkage to poor nonword repetition skill was found for a locus on chromosome 16 and was replicated across a second sample. (Another linkage, to a site on chromosome 19, was found for a more general expressive language measure.) It is important to emphasise that finding linkage is not the same as finding a gene; there are likely to be hundreds of genes close to a linkage site. Furthermore, when we move away from clearcut pedigrees such as that of the KE family, the task of locating a relevant gene is much more difficult, because we are looking for probabilistic associations between an allele and a disorder, rather than a one-to-one relationship. Nevertheless, the linkage study has narrowed down the field of search, and has also demonstrated the potential value of using theoretically motivated language indices, rather than more conventional clinical criteria, for defining disorder.

Language as a continuous trait

If we find a gene that is associated with nonword repetition ability, it will not necessarily be like FOXP2, where a simple mutation leads to clearcut language impairment. Many complex heritable disorders are *polygenic*: that is, they are the result of the combined influence of many genes, each of small effect. Furthermore, the alleles responsible for causing variation may be common in the normal population, unlike the extremely rare FOXP2 mutation. Many of my colleagues in this field think it is misguided to suppose that we will find many other single genes that have a large effect on language when mutated. Instead, they argue, we should view SLI not as a disease, but as an extreme point on a normal distribution of ability (see Figure 7.5). We may draw an analogy with height. If we select from the population those people whose height is well below average, we are likely to have a mixture of cases. Some will be people whose short stature is due to a recognised medical condition – possibly a rare mutation of a single gene. However, many cases of short stature have no obvious medical cause. Rather, they fall on the extreme of normal variation. We know that height in the general population is influenced by a number of polymorphic genes, each of which exerts a small effect. Environmental factors, such as diet, are also important. If we study people of short stature, we will be able to see that there are substantial heritable influences on height, but we are unlikely to find a single mutated 'height gene'. We are even less likely to find a gene that differs between human and chimpanzee, even though the two species differ substantially in average height. Behaviour geneticists such as Robert Plomin have argued that we should expect a trait such as language to behave in the same way, and that SLI will turn out to be the consequence of additive influences of many genes of small effect. It seems likely that a great

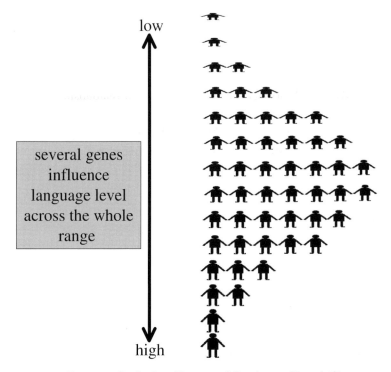

FIGURE 7.5 Frequency distribution of language ability that would result if language level were determined by many genes each of small effect. Low language ability would be the consequence of combined action of many alleles, rather than the result of a single mutated gene. When a trait is affected by multiple small influences, we see the classic normal distribution, or 'bell curve', with most people obtaining scores in the middle region of the distribution, and smaller numbers scoring at either extreme.

deal of variation in language abilities in young children is related to individual differences in rather general processes that may affect, for instance, the rate of brain maturation, rather than having selective effects on language.

If SLI is just the tail end of a normal continuum, the study of SLI will not get us much further in our quest to account for the distinctively human language capacity. My own view, however, is that this is too restricted a viewpoint. It is likely that a polygenic mechanism will account for a proportion of children with slow language development, and the kind of arguments advanced by Plomin provide a salutary message for anyone who thinks that the existence of heritable SLI is *proof* that there are language-specific genes. However, we still need to explain why humans learn language easily and chimps fail to do so. The notion that there may be several human-specific genes that are important for language is, in fact, far more plausible in evolutionary terms than the original 'hopeful monster' notion. Language is a complex system of interconnected functions which has considerable survival value. One may be tempted to think that such a complex system, which involves interconnected hierarchical representations of sounds, meanings and grammatical structures, could not possibly have evolved. However, this line of reasoning has the same flawed logic as the arguments of Victorian clerics who maintained that a complex structure such as the eye had to be the work of a divine creator, as it could not be explained by evolution. In fact, as Steven Pinker has argued, evolution by natural selection is the only tenable scientific account of the emergence of complex functional systems. Derek Bickerton was the first person to make a serious attempt to characterise proto-language – an intermediate stage between no language and full-blown syntactic competence. More recently, another linguist, Ray Jackendoff, has speculated on the different stages that language may have gone through in the course of evolution, and proposed no less than nine steps between the ability to use basic symbols and modern human language. Neither Bickerton nor Jackendoff talks in terms of innate knowledge of rules of grammar; rather they propose the existence of specialised brain systems that are set up to detect particular types of patterning in linguistic input. Furthermore they argue that we would not expect these specialised systems to arise out of thin air; just as a swim bladder evolved into a lung, we would expect language processors to evolve out of brain systems that originally carried out other functions. Insofar as such processes exist, and are under genetic influence, then it follows that we might find people in whom they are defective because one of the relevant genes is mutated.

Another reason for persisting in the hunt for language-specific genetic influences is that there are some aspects of language that *do not* behave like height, insofar as there is very little variation from person to person. Individual variation is clearly apparent in the rate of language acquisition, and it is likely that such differences in maturation do depend, not on language-specific mechanisms, but on much more general genetic influences. Furthermore, normal distributions of language scores are seen for tests that assess such verbal functions as vocabulary size, reasoning, or working memory capacity. These are abilities where it is reasonable to postulate a role for polygenic influences

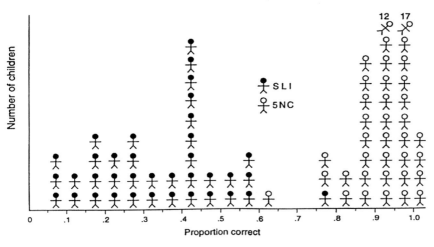

FIGURE 7.6 Distributions of scores obtained on a test that assessed children's ability to add inflectional endings to verbs. SLI group shown in black, typically developing 5-year-olds (5NC) shown in white (reprinted, with permission, from: M. L. Rice, 'Grammatical symptoms of specific language impairment', in Bishop and Leonard (eds.), *Speech and Language Impairments in Children: Causes, Characteristics, Intervention and Outcome*).

similar to those that operate for general intelligence. However, as Chomsky emphasised, other aspects of language show remarkably little variation once they are mature. Syntactic skills are pretty uniform in adults, and are mastered by most children by 4 years of age. Likewise, all normal humans with adequate hearing quickly develop the ability to perceive a continuously varying acoustic signal – speech – in terms of a small number of categorical units – phonemes – which themselves are meaningless but can be combined to form meaningful units – words. Problems with these aspects of language formation are unusual, and when they do occur are recognised as clinically significant, warranting a diagnosis such as SLI. Illustrative data from a study by Mabel Rice are shown in Figure 7.6. The ability of young children to produce inflectional endings is not a normally varying trait – most children do this impeccably by the age of 5, but a sub-set – those with SLI – have significant problems. There is little overlap between the two distributions.

If we are going to have any success in our hunt for language-specific genes, it makes sense to focus more on those distinctive aspects of language structure that show little individual variation in the general population, rather than more global measures of language function that are normally distributed. Work on this topic is only just beginning, but I expect we may see a different pattern of genetic influence on SLI if we focus on language traits that behave this way, rather than relying on conventional language tests.

Conclusions

During the latter half of the twentieth century, debate on the issue of language specialisation in humans tended to be polarised between linguists who appeared to finesse the problem by postulating an innate language module, and psychologists who denied that there was anything special to explain. As we move into the twenty-first century, a number of more moderate voices are beginning to be heard from those who see the question not as 'do humans have innate grammar?' but as 'how far does language acquisition require specialised brain mechanisms, and what are they?' Genetic studies of children with language impairments will not provide all the answers to such questions, but they provide a unique source of data that has the potential to illuminate our understanding of the complex pathways from genes, through neurobiology, to behaviour.

FURTHER READING

D. V. M. Bishop, *Uncommon Understanding: Development and Disorders of Language Comprehension in Children*, Hove: Psychology Press, 1997.
  'The role of genes in the etiology of specific language impairment', *Journal of Communication Disorders* **35**, 2002, 311–28.

W. H. Calvin and D. Bickerton, *Lingua ex Machina: Reconciling Darwin and Chomsky with the Human Brain*, Cambridge, MA: MIT Press, 2000.

D. L. Cheney and R. M. Seyfarth, *How Monkeys See the World*. Chicago: University of Chicago Press, 1990.

J. Elman, E. Bates, M. H. Johnson, A. Karmiloff-Smith, D. Parisi and K. Plunkett, *Rethinking Innateness: A Connectionist Perspective on Development*, Cambridge, MA: MIT Press, 1996.

R. Jackendoff, *Foundations of Language*. Oxford: Oxford University Press, 2002. (recommended only for those with some background in linguistics).

G. E. Marcus and S. E. Fisher, 'FOXP2 and the search for "language genes"', *Trends in Cognitive Sciences* **7**, 2003, 257–62.

S. Pinker, *The Language Instinct: the New Science of Language and Mind*, London: Penguin Books, 1994.

R. Plomin and J. Crabbe, 'DNA', *Psychological Bulletin* **126**, 2000, 806–28.

S. Savage-Rumbaugh, S. G. Shanker and T. J. Taylor, *Apes, Language, and the Human Mind*, New York: Oxford University Press, 1998.

M. Tomasello, 'Acquiring syntax is not what you think'. In D. V. M. Bishop and L. B. Leonard (eds.), *Speech and Language Impairments in Children: Causes, Characteristics, Intervention and Outcome*, Hove: Psychology Press, 2000.

J. Weiner, *Time, Love and Memory*, London: Faber & Faber, 1999.

8 DNA and ethics

ONORA O'NEILL

Newnham College, Cambridge

We are not short of public discussion of ethical issues arising from genetic knowledge. Enthusiasts and scaremongers fill the columns and the airwaves with their hopes and fears about 'designer babies', GM crops, pharmacogenetics and a future in which everyone carries a genetic smart card. Each of these possibilities, like any other complicated change, would raise ethical issues. The hardest ethical issues that we currently face, however, are rather different. Many have to do with the use and control of genetic information. Is genetic information exceptional? Are there good reasons to control access to it more tightly than we control access to other sorts of personal or medical information? Does genetic information 'belong' to individuals or to families? Could there be a 'right to know' the results of DNA tests taken by relatives, or a 'right not to know' – or both? *How* informed do we have to be to give informed consent to genetic tests?

Enthusiasts and scaremongers

Few public debates are more polarised than those about new genetic technologies. Yet the enthusiasts and the scaremongers agree on one point. Both take it that the most serious ethical problems cluster around possible, or likely, modifications of living organisms by new genetic technologies. Of course, they disagree radically in their assessments of the opportunities and risks that these technologies present, and about what we should and should not do.

Enthusiasts think that the use of technologies for genetic modification will generally prove liberating and beneficial, especially for human beings: we are entering a brave new world. Medicine, in particular reproductive medicine, will be transformed: we shall cure illness with genetic therapies and research in pharmacogenetics will lead to better targeting of medicines. We shall have genetically 'enhanced' children, with chosen and desirable traits. Agriculture will be transformed by genetically modified plants and animals, which will go far to solve the problem of world hunger, reduce use of pesticides, provide medicines and save the environment.

Scaremongers think that using technologies for genetic modification will harm and destroy human and other organisms, and their environment: they point to doomsday scenarios. They warn us that genetically modified organisms – bacteria or plants, animals or humans – might have unforeseen defects, that they might proliferate or hybridise, create unforeseen plagues or environmental damage, and might divide human societies into a permanently privileged group of the 'genetically enhanced' and a 'genetic underclass'.

These disagreements colour views on public policy. Enthusiasts often take a libertarian stance. For example, John Harris of the University of Manchester has argued for extensive reproductive *autonomy*, including a supposed 'right to reproduce with the genes we choose and to which we have legitimate access ... in ways that express our reproductive choices and our vision for the sorts of people we think it right to create'.[1] Scaremongers often take a prohibitionist stance. For example, the Greenpeace briefing on GM crops opposes the release of all and any GM organisms into the environment.[2] Enthusiasts and scaremongers share a tendency to sensationalise, to glamorise or to demonise, genetic technologies, rather than to judge applications in the light of evidence. It is hardly surprising that they are both drawn to quite extreme ethical positions.

At the risk of being much more boring, I want to discuss some ways in which ethical questions raised by uses of genetic technologies are more ordinary, more diverse, and demand more detailed consideration of specific applications and their likely effects than some enthusiasts or scaremongers think necessary. I shall start with technologies for genetic modification of human and other organisms, and will then turn to technologies for using genetic information.

---

[1] John Harris, 'Rights and reproductive choice', in John Harris and Søren Holm (eds.), *The Future of Human Reproduction: Ethics, Choice and Regulation*, Oxford: Clarendon Press, 1998, pp. 5–37; p. 34.

[2] *Genetic Pollution – a Multiplying Nightmare*, Greenpeace, February 2002.

Genetic modification: human health and reproduction

Consider for a moment where we *actually* stand in the application technologies for genetic modification to human beings, and in particular what gene therapy and a range of reprogenetic technologies currently offer.

Gene therapy is still at a very early stage – although there are many non-genetic medical interventions for managing, and some for preventing, genetic diseases. Haemophilia can be managed; the retardation to which Phenyl-ketonuria (PKU) can lead can be prevented. Still, cure is better than treatment, and this is what gene therapies might provide. However, at present (with few exceptions) gene therapy is at the stage of research or clinical trials. These trials are closely controlled by the Gene Therapy Advisory Committee (GTAC) in the UK (and by parallel bodies overseas); in the UK germ line gene therapy, where modifications would be transferred to future generations, is prohibited. If and when reliable gene therapies are developed, patients may not find them very different from more ordinary medical interventions.

Reprogenetics at present also offers pretty limited new possibilities: yet discussion of the possibility of having genetically 'enhanced' children ('designer babies') gallops ahead of reality. At present genetic technologies are just not very useful if you want to design a baby. Couples who know that they risk having children with an identifiable single gene disorder – say, cystic fibrosis or Tay Sachs disease – can now choose not to do so, either by using in-vitro fertilisation (IVF) with preimplantation diagnosis (very rare), or by post-conception genetic testing and abortion if the foetus is affected (less rare). In the past such couples had fewer options: they had to choose between childlessness and adoption, or risk having an affected child. Using the new technologies to avoid having a child with a specific, anticipated and very damaging genetic variation is a far cry from designing a child. The reprogenetic technologies in *actual* use extend but do not revolutionise parental choice, and the ethical problems they raise are hard but familiar.

Similarly with technologies for *sex selection*. Sex selection can be done (to a degree) using genetic and other new technologies, and is strictly regulated in the UK. Advocates of more liberal access to new technologies for sex selection see it as important for reproductive choice; opponents point to the unintended consequences, typically a large excess of men (not of women). These unintended consequences have arisen, particularly in India and China, where sex selection is routinely secured on a massive scale by traditional methods such as neglect, abandonment and infanticide of little girls, and (more recently) by amniocentesis and selective termination. Some technologies may be new, but the issues are not.

This brings me back to *designer babies*. There are at present *no* genetic technologies for choosing most of the characteristics that parents prize, such as intelligence or a happy disposition. To the best of our knowledge there are genes that contribute to the likelihood of developing these and many other desirable (and undesirable) traits, but no single gene is decisive. If single genes were decisive, we would probably have discovered their pattern of inheritance long since, as we did for single gene diseases such as Huntington's, cystic fibrosis or haemophilia. Since we do not yet know much about the genetics of polygenic traits, about interactions between susceptibility genes, or about gene–environment interactions, we do not yet know what would have to be done to design children with these prized human characteristics.

If we do gain that knowledge, designing parents would face the risks and rigours of IVF combined with complex genetic modification (as opposed to embryo selection). At present we are many steps away from the designer baby scenario, lacking adequate knowledge of susceptibility genes or their interactions, and lacking acceptable and reliable techniques for complex genetic modification of human embryos. For the time being, and I take it for a good time to come, a good education and a cheerful home remain the best bet for those who want an intelligent, happy child. What is on offer is simply less promising, or less threatening, than many discussions suggest, and we can do no more than ensure that we maintain robust ways of controlling the application of any further reprogenetic technologies.

Genetic modification: micro-organisms, plants and animals

Enthusiasts and scaremongers focus on present as well as possible future scenarios when they discuss technologies for the genetic modification of micro-organisms, plants and non-human animals. These technologies are neither as alien nor as novel as some scaremongers often suggest, nor as transformative as some enthusiasts hope. Genetic modification is not a recent or unheralded innovation. It is the engine of evolution and the basis for traditional and not-so-traditional plant and animal breeding. Perhaps the most appropriate way to judge the acceptability of uses of GM technologies is to compare them with earlier forms of genetic modification.

Let me start with micro-organisms. Many have a talent for genetic modification. Bacteria achieve antibiotic resistance by picking up novel DNA from other bacteria; the influenza virus changes itself so readily that it makes vaccines obsolete, and causes significant human illness and mortality. Technologies for genetic modification have been applied to micro-organisms since the 1970s. Many modifications have been permitted, and proved valuable. Most vitamins, antibiotics and amino acids are produced in genetically modified bacteria. Genetically modified bacteria are used in making 85% of the cheese eaten in the UK, in place of rennet, taken from the stomachs of calves – a change that is advantageous to vegetarians, as well as to calves. Micro-organisms modified to produce human insulin and human growth hormone are routinely used and have produced considerable health and safety benefits. Other genetic modifications of micro-organisms might prove useless or dangerous. Modified bacteria and viruses *could* be used to spread human and animal disease, as recurrent worries about bio-terrorism remind us. Given the real benefits and possible harms produced by modified and unmodified micro-organisms, the only reasonable approach is to maintain systems for detailed, effective regulation backed by powers to prohibit research and use of modified micro-organisms whenever there are reasons to do so.

Similar things might be said about the genetic modification of plants, which is the scene of so much controversy in the UK and the EU. Genetic modification of plants takes place entirely naturally, and not always to the benefit of human and animal life: the world is full of successful plants that achieved genetic modifications that made them taste nasty and have toxic effects on various animals and insects. The world is also full of plants that hybridise promiscuously, sometimes with mildly undesirable results: the wheat and tares of the Bible are *both* of them genetically modified plants. The wheat was produced by human manipulation, by the action of generations of Stone Age men and women, and has provided the staff of life for countless millions. The tares were a back-cross between domesticated wheat and its own wild ancestor, einkorn wheat. Tares are a problem for farmers – but not a catastrophe.

So the wholly unspecific question 'Are genetically modified plants a good or a bad thing?' is unlikely to have a sensible answer. Specific modifications of particular plants may very well turn out to be beneficial or harmful in various ways for various species, or neither (traditional plant breeders find that few of the genetic modifications they produce are of the slightest value). Whether a modification is harmful or valuable seems to me independent of the way it was produced. Modifications produced either by traditional technologies or in the laboratory *might* be useful, useless or harmful. The sheer variety of possible outcomes is a good reason to regulate research on genetically modified plants, release into the environment, and agricultural use with care and rigour in the light of evidence – and good reason to apply equally tough standards to traditional plant breeding and the importation of exotic plants.

Prohibitionists think a selective, evidence-based approach to GM plants inadequate. Some suggest that the so-called 'Precautionary Principle' offers a conclusive reason for prohibiting *all* laboratory-based genetic modification of plants. There are many formulations of this principle, and some of them do not offer a plausible or useful guide to action. When the Precautionary Principle is seen as a requirement to take care, to be prudent and cautious, it is very plausible. An injunction to be prudent is not precise, or thrilling; but it is ethically important. Adequate regulation of genetic technologies has surely to be based *in part* on a commonsense interpretation of the Precautionary Principle, that takes serious account of evidence and is combined with due attention to other ethically important considerations. How else could anyone judge which acts are cautious, rash or neither?

However, prohibitionists do not think commonsense interpretations of the Precautionary Principle adequate. They recommend supposedly stronger versions, which (they claim) provide reasons for avoiding all GM technologies, indeed all new technologies, that *might* have bad consequences. The stronger versions of the principles are now sometimes formulated (in spurious homage to scientific method) by saying that we must take account of so-called 'unknown unknowns', or in more politicised versions by a general claim that we should avoid 'possible risks' or 'shift the burden of proof to those who create risk'.[3] Unfortunately, far from providing a knock-down argument against any planting of GM crops, or against using other new genetic technologies, these versions of the Precautionary Principle cannot help anyone to decide which technologies to use and which to shun.

I believe that the intellectual credibility of the Green movement has been weakened by its enduring love-affair with supposedly strong versions of the Precautionary Principle. Their fatal attraction is an illusory strength, tempting

---

[3] John Humphries, *The Great Food Gamble*, London: Hodder and Stoughton, 2002, p. 109.

because it suggests a way of bypassing the chore of assembling and assessing the evidence for and against proposed courses of action. The strong formulations of the principle seem attractive because they can be cited as reason to reject *any* GM crop, since it *might* have some bad effects. Yet the same versions of the principle also torpedo all the innovations that Greens advocate (exclusive reliance on biological control of pests and banning artificial fertilisers *might* have bad effects). They also undermine staying with the status quo: there is little doubt that intensive farming *actually* has various bad (as well as beneficial) effects. Serious arguments for and against GM crops, like serious arguments for or against organic farming, or for or against intensive agriculture, must take account of the evidence for or against *specific uses* of *specific technologies*. A claim that we should avoid all possible risks or 'shift the burden of proof to those who create risk' cuts no ice unless we can identify some line of action that creates zero risk. Since we cannot do that, we need to consider *which* technologies would create *which* risks, and which risks are *more* and which *less* tolerable; but we cannot do this without taking a rigorously evidence-based view of specific uses of specific technologies.

I suspect that most Greens have rather less faith in the Precautionary Principle than some proclaim. If they *really* thought that they had a knock-down argument against new technologies, including GM technologies, they would view all appeals either to evidence or to other ethical principles as redundant, and do without them. Yet they in fact constantly appeal to selective evidence, for example to the benefits of compost, or of biological pest control; and they often appeal to other ethical principles, such as the importance of limiting human and animal suffering and (various, supposed) human rights and benefits. More startlingly, they quite often appeal to a libertarian principle of consumer choice: this principle is supposed to underpin a requirement to label GM products. Those who invoke it may find that it boomerangs, since it suggests that consumers have a right not only to know which products contain GM ingredients, but also to have such products on offer.

Since the supposedly strong versions of the Precautionary Principle succumb to this *reductio ad absurdum*, we can gain more from the weaker versions. These we have reason to take entirely seriously in working out *which* uses of *which* technologies for genetic modification there is reason to ban or to permit, under *which* conditions and for *which* purposes. The devil, inevitably, is in the detail.

The problem is to combine prudence and caution with other ethical concerns, such as limiting human and animal suffering, improving food security for the poor, securing greater justice and benefiting health and welfare. Those who take the trouble to think through the details may or may not reach what are currently seen as Green-ish conclusions. Full consideration of the ethical issues and of the evidence *might* endorse organic farming plus severe restriction or prohibition of GM. Alternatively it *might* endorse a combination of Green-ish and high-tech approaches – for example organic farming with GM crops; or it *might* point in other directions.

Similar considerations are relevant to genetic modification of animals. There is no doubt that some genetic modification of animals, produced by traditional as well as by new methods, may cause harm and suffering. Think how much animal suffering the zealous breeders of dogs with a 'slinky' look (and hip dysplasia) or squashed faces (and breathing problems) have caused. They failed to think about the whole animal, and bred for isolated 'features' or 'points'. Other genetic modifications – whether achieved by traditional or by newer techniques – may benefit not only the species modified, but also other animals (including humans) or the environment. For example, genetic modification of sheep *may* eliminate susceptibility to scrapie, presumably with benefit to sheep, and to other animals, including humans. Genetic modifications of pigs *may* reduce their excretion of methane and thereby their contribution to greenhouse gases, so benefiting the environment. The assessment of these possibilities does not depend on whether modifications arise naturally, are produced by selective breeding or are the result of genetic modification in the laboratory.

Once again the strong versions of the Precautionary Principle cut no ice. A commonsense version, recommending caution in all we do, is well taken; but caution is illusory unless it is evidence-based, and is only one of many ethical considerations relevant to the complex decisions that farmers and breeders, scientists and policy-makers, and the public at large have to make. The exorbitant versions of the Precautionary Principle that prohibit action unless it is *guaranteed* to have no bad effects can only paralyse. We can be cautious, we can even choose to 'err on the side of caution'; but this requires rigorous consideration of cases and evidence-based assessment of new and traditional technologies. Prudence and caution are undermined rather than brought to perfection unless we take evidence seriously.

Genetic information: identity and sense of identity

So far I have argued that technologies for genetic modification are often less dramatic and less strange than enthusiasts and scaremongers suppose, and that the ethical problems they raise are more similar to those raised by traditional technologies than they assume. We live in a world in which genetic therapies and genetic modifications have to be evaluated and used with rigour and care, in the light of relevant evidence. There is no *single* or *simple* principle for deciding which technologies to prohibit, or what conditions to set on those not prohibited. There is no way to avoid the hard work of assessing possibilities and cases as carefully as we can, in the light of available evidence and of robust rather than gestural ethical arguments.

Readers may by now suspect that a discussion of Ethics *and DNA* is redundant. Yet a look at genetic technologies that are *not* used to modify organisms shows that some of them give rise to quite distinctive ethical problems. The genetic technologies to which I shall now turn are technologies that deploy *genetic information that pertains to individuals*. These technologies have already been put to many uses.

Genetic information – mainly DNA information – obtained from individuals is used for many medical purposes other than gene therapy and genetic modification. It can play a role in diagnosis and in treatment, and can inform reproductive decisions. If pharmacogenetics develops, it will be used routinely in prescribing medicines. Genetic information obtained from individuals can also be linked to other information about the same or about distinct individuals. Technologies for handling and linking DNA information are already in daily use by the police, the immigration service, the child support agency, family historians, archaeologists, stockbreeders and many others, and are important for public health and health research.

Genetic information that pertains to individuals is widely seen – not only by enthusiasts and scaremongers – as distinctive, personal and peculiarly sensitive, even as vital to our very identity. The phrase 'genetic identity' has acquired a wide currency, and has migrated from academic to popular discourse. A typical example in the *Independent* in September 2000[4] quotes a man conceived by donor insemination as saying that not knowing anything about his biological father leaves him feeling that he is 'missing 50% of my genetic identity'.

---

[4] Marie Woolf, 'New Rights for Children of Sperm Donors', *Independent*, 19 September 2000.

The term 'genetic identity' sounds as if it refers to something important, even foreboding; yet its current uses often refer to, and confuse, a number of quite distinct matters.

On a first, metaphysical, understanding of the notion of *genetic identity*, genetic makeup is viewed as the basis of personal identity, as that which makes people distinct from one another and reveals what is fundamental about them. This conception of genetic identity is criticised by some academic writers as implying some form of *genetic essentialism*, but is quite often endorsed in more popular (indeed populist) writing on genetics, as exemplified by the catch-phrase 'genes are us'.

The claim that genes are the basis of the identity of persons is clearly false. Genetic makeup is not sufficient to individuate a person: identical twins who share their genetic makeup are distinct persons. They differ in many other characteristics, initially in gestational history and time of birth, thereafter in countless other properties. The fact that genetic difference would *generally* be sufficient to distinguish one individual from another is irrelevant. Precise time of birth is *generally* sufficient to distinguish one individual from another; but we know that 'birth time identity' is not always sufficient to individuate persons.

On another, cultural, interpretation of the notion, *genetic identity* does not mean that individuals are *distinct*, but rather that they *share* something. During the last two decades phrases like 'ethnic identity' or 'religious identity' have been used not to pick out *distinct* individuals, but to pick what individuals *perceive* or *represent* themselves as *sharing* with others. Knowing a person's social, national or religious identity – or (as we used more accurately to say) their *sense of identity* – amounts to knowing what they share with certain others. If my ethnic identity is Cornish or Kurdish this will show whom I view as fellow countryman and whom as outsider; if my religious identity is Coptic or Catholic this will show whom I view as fellow communicant and whom as heretic, or at least religiously misguided. *Senses of identity* are not generally matters of discovery, but rather of known identification, commitment and aspiration; they are clusters of shared beliefs, aspirations and ideals. I suspect that the person who felt that ignorance of his father damaged his 'genetic identity' meant that he lacked knowledge of things that he shared with his father and his paternal lineage.

Yet I do not think that the knowledge he craved is best described as knowledge of his *genetic identity*. What he sought was knowledge about his family tree, that is knowledge of lineage and origins, and would be better thought of as knowledge of *genealogical identity*, or more prosaically as *genealogical information*. Gaining this knowledge would not yield much (if anything) in the way of knowledge of specific genetic variations that he shared with his paternal lineage. It would tell him *that* he shares genes with certain others, but not *which* genes he shares. This may not matter much, given that beliefs about one's genetic makeup are not (usually) important to anyone's sense of identity. Most of us have minimal beliefs about or commitments to our genetic profile or its implications, but this does not weaken our sense(s) of identity. The term 'genetic identity' misleads because it suggests that genealogical information will of *itself* enrich, enlarge or contribute to a person's sense(s) of identity. In fortunate cases those who gain new genealogical information may identify with previously unknown relatives, with whom they come to share certain beliefs or aspirations: their sense(s) of identity may converge with those of their newly identified relatives. In other cases there will be no happy convergence of sense of identity.

Personal genetic information: sharing and anonymity

If genetic information that pertains to individuals is *neither* the basis of personal identity *nor* the basis of sense(s) of identity, is it of any particular ethical importance? Should it be treated differently from other personal information? Is it any more sensitive than the personal information found on drivers' licences and passports?

Those who believe that genetic information is exceptional often suggest that this is because it is *familial* rather than (solely or strictly) *individual*. It is true enough that genetic information is familial. But what follows? Many other sorts of personal information are familial, such as information about family fortunes or family jokes. And even if some familial information is sensitive and something we seek to keep private, not all of it is. Families may not want information about their quarrels and their debts to be public, but may hope that information about their achievements and distinctions is public. Indeed, some familial information is unavoidably public: it is hard to hide prominent family traits like baldness or the Habsburg nose.

So a *general* claim that genetic information *is* exceptional or peculiarly sensitive simply because it is familial seems ill founded. Nevertheless certain *uses* of genetic information, and in particular of DNA information, may raise distinctive ethical problems. I believe that these problems cluster where the information is not obvious (so not unavoidably public), yet pertains to more than one individual. In discovering such information about ourselves we may also discover something nonobvious about our relatives, possibly something that they do not know about themselves. These small facts give rise to ethical quandaries because they challenge current individualistic conceptions of personal information, personal privacy and informed consent.

The most obvious cases arise where individuals have DNA tests in the course of medical treatment or for reproductive purposes. What makes the test results sensitive is their nonobvious medical and reproductive implications; what makes them distinctive is that (unlike most such information) they also pertain to relatives. For example, a person whose grandparent died of Huntington's disease and whose parent refused tests may choose to be tested and discover that he or she has inherited the gene, so will suffer the disease, and thereby discovers that the parent who refused testing unknowingly has the same prospects (matters are quite definite here, as in few other cases, because the Huntington's gene is dominant and highly penetrant). More typically, people whose DNA test results reveal genetic risks can infer that certain relatives are also at risk. They obtain sensitive and difficult information about their relatives. Disclosure to those relatives might prompt radical changes in reproductive decisions and life plans; nondisclosure is equally morally problematic.

The ethical puzzles that arise are most easily stated for the case of identical twins. In seeking DNA information about myself – information that is presumed on a standard, individualistic view to be mine to seek – I discover information that is as true of my identical twin as of me. Does it follow that my twin had a right to insist that I seek prior consent, or that my twin's refusal would have limited my right to seek such information about myself? And once I obtain DNA information about myself, does my twin have a right to share the knowledge? Or, alternatively, a right not to share it? What practical content and what arguments could be given for any of these supposed rights? If DNA tests required consent from (all possibly at-risk members of) families, would individuals have to do without tests, even if medically important, whenever one or another relative refuses consent? And what is to be done where relatives cannot be contacted or disagree? Is there a right to 'genetic privacy', and what does it cover? Do relatives have rights that limit one another's rights to privacy?

177

As you can see, these issues are often discussed in terms of appeals to supposed rights. I suspect that this proliferation of problematic and unargued rights claims means that we have spotted some interesting questions, but remain pretty much in the dark about answers. In general we do not gain much clarity about rights claims until we can formulate the counterpart claims about duties, and in this area we have hardly begun. Just listing some of the duties that would correspond to these supposed rights is revealing. If relatives-at-risk have a *right to know* DNA information obtained by an individual, then that individual has a *duty to share* the information obtained, hence only a restricted *right to personal privacy*. On the other hand, if relatives-at-risk have a *right not to know*, then an individual who learns difficult or upsetting DNA test results has a *duty not to share* this information, unless the relatives waive their rights. Yet how could relatives-at-risk who do not know what has been discovered waive either a *right to know* or a *right not to know*? And if relatives-at-risk have both a *right to know* and a *right not to know* genetic information that pertains to them, from individuals who obtain it, then that individual would have both *a duty to share* and *a duty not to share* the same information with the same relatives-at-risk. Unrestricted versions of the two rights evidently conflict; but which restrictions are convincing?

These puzzles suggest that we have barely begun to articulate the challenges that certain uses of genetic (and perhaps other) information pose to individualistic conceptions of personal information, privacy and informed consent. Nor are matters made easier if we abandon the vocabulary of rights and try to articulate the issues in terms of informed consent requirements. In the first place, such requirements are themselves often supposedly justified by reference to rights (questionable conceptions of individual autonomy sometimes playing a mediating role). Secondly, informed consent requirements prohibit acting on others in ways to which they have not freely consented on the basis of (full) information. Thirdly, they prohibit use or disclosure of personal information about others except where those others have freely consented to use or disclosure. Yet we often permit individuals to use and disclose information about others without their consent. For example, the hallowed medical practice of taking a family history invites patients to disclose information about the health of relatives *without their consent*.

Is the project of classifying information exclusively and exhaustively *either* as personal to individuals (so subject to informed consent requirements) *or* as public (so not subject to such requirements) perhaps misguided? Or are these puzzles raised by familial DNA and health information atypical? Perhaps such cases are confined to clinical genetics and reproductive medicine, where account can be taken of the problems created by information that is accessible to individuals yet does not pertain solely to individuals. Perhaps other uses of DNA information are just less problematic. After all, in many cases there is no need to establish *any* of the medically or reproductively sensitive implications of a person's DNA. For example, when DNA information is used for forensic purposes, samples are simply matched and their health and familial implications need not be established, communicated to the affected individuals, or disclosed to relatives, let alone made public.

Matching DNA information is perhaps just like matching other sorts of evidence. This was certainly suggested by the title of a 1999 Home Office consultation, which linked the rather surprising trio of topics 'Footprints, fingerprints and DNA samples'. All three types of information are forensically important, because samples from different sources can be matched. By itself, a discovery that two DNA samples – or for that matter two footprints – match cannot identify a person. The discovery of matching samples of DNA is, however, reliable evidence that one and the same individual (or, of course, his or her identical twin!) was the source of both samples. Given other information, a DNA match may show that a single individual was present at two scenes of crime, and ultimately provide a basis for arrest and conviction. Equally, given other information, the lack of a DNA match can eliminate suspects or suggest that distinct individuals were present at two scenes of crime. Matched DNA samples do not by themselves identify persons – let alone prove whodunit. Once linked to other information, however, they are of high evidential value. Footprint evidence is rather less useful: a match may lead one to someone else with the same shoes, and is useless if shoes are thrown away. Even matching fingerprints may be smudged, and *in extremis* skin grafts can undermine their evidential value. By contrast, the evidence provided by DNA samples retains its value through and far beyond individual lives.

In forensic uses of DNA matching, issues about the sensitivity of genetic information apparently vanish because (in general) nothing sensitive is determined or revealed. The police have no interest in – and no budget for – finding out whether a suspect carries a gene for sickle cell disease or for familial forms of breast cancer. They do not seek out or interpret the medical information that is latent in the DNA samples that they match. They use the information to get their man or woman – and to eliminate irrelevant suspects. The ethical issues that arise in the use of DNA samples for forensic purposes are therefore quite different from those that arise in considering GM technologies: they centre on the use of state power to request and retain samples, and on the dangers of contaminated evidence and of unauthorised uses of a forensic database. They have nothing, however, to do with the sensitivity of information about health or family prospects, or (excepting the identical twins) with the fact that genetic information is not strictly individual.

Nevertheless, other uses of technologies for matching DNA samples raise questions that are as sensitive as those that arise in medical and reproductive contexts. The matching techniques that are used to determine whether samples come from a *single* individual can also be used to determine whether they come from *related* individuals. Genetic profiles can be used to settle paternity or nonpaternity, to link lost children to their relatives, to test the kinship claims of would-be immigrants and would-be inheritors, and to establish kinship with unidentified human remains. DNA matching may raise or resolve claims in some of the most fraught and distressing situations that we can imagine, many of them familial situations. These typical uses of DNA matching raise both the questions about authorisation and proper use that arise in forensic uses, and those about the limits of individual and familial claims that arise in medical and reproductive contexts. Once again, it is because matching is with putative relatives, because DNA information is not solely individual, that ethical problems arise. Once again, a view that informed consent by an individual legitimates DNA matching is placed under great pressure. Indeed, DNA matching can be undertaken without consent from those most affected, or from their relatives: third parties have only to lay hands on a hair or tissue and a website, and to pay the fee.

The difficulties may lie even deeper. The deeper problems may be not that there is a mismatch between an individualistic conception of informed consent and the reality that DNA information is familial. It may be that the underlying assumption that information is always private to individuals of whom it is true should be questioned. I have already noted that we regard a great deal of obvious information that has a genetic basis – e.g. hair colour and eye colour – as routinely and unavoidably public. But when and why is nonobvious information to be seen as private to those of whom it is true? We cannot answer this question by appealing to an unargued notion of 'genetic privacy': the point is to determine what is and what is not private.

The importance of thinking through these elementary issues is becoming more urgent with the establishment of databases that link DNA with other types of information about individuals, such as information about health, medical treatment or lifestyle, and even genealogical information. Such databases – the DeCode project in Iceland, prospectively BioBank in the UK – are seen as important for health research. Is the storage, processing and use of data acceptable provided that those to whom the information pertains have given their informed consent? Is consent adequately informed if it conforms to data protection requirements?

Here I think we may feel pretty uneasy. Data protection legislation takes no account of the possibility that some information may not in the first place pertain just to individuals. In fact, it takes a more radically individualistic view of personal data and informed consent than has been traditional in medical practice, or elsewhere. In the past, health research was deemed ethically acceptable provided that adequate anonymisation prevented unauthorised identification of individual patients beyond the circles of those responsible for their treatment. Current standards of data protection demand more. They demand *either* that all information that pertains to individuals be irreversibly anonymised (which precludes data linkage) *or* that those who use or link the data seek informed consent from the individuals to whom it pertains for the specific uses to which the data are put.

This has created a crisis for any retrospective use of medical information, or DNA information, for health research that was not (and often could not be) anticipated when the information was first collected. If prior consent from data subjects were required, information on former patients could never be used in investigating new diseases. It is a fantasy to imagine that prior consent can be given to future research projects.

Indeed, it is unclear whether *any* ethically convincing form of informed consent to highly complex uses of DNA information is possible. Individuals may reasonably feel that their capacities to grasp information and to give or refuse informed consent are simply overwhelmed if they are asked to consent to inclusion of very complex information in massive data sets linked in complex ways with other complex data. Yet if data linkage without specific consent from data subjects is unacceptable, and data linkage with their consent is unattainable, how are we to maintain public health records and how can we conduct health research? More generally, how can we make use of knowledge gained from treating one patient in treating others? Given that all patients are treated on the basis of medical knowledge of previous patients, can it make sense to subject all use of information about any individual patient to his or her consent? Or might we have reason to reconsider data protection requirements and revert to the view that anonymisation – including reversible anonymisation – is adequate protection of individual privacy? How might we take better account of the fact that some information is not solely individual, of the reality that capacities to understand information are limited, and of the reality that health policy has to draw on information that pertains to individuals even when explicit consent cannot be obtained? These are large and topical questions which need a lot of attention. They are good reasons for looking beyond current battles about genetic modification.

FURTHER READING

Michael J. Reiss and Roger Straughan, *Improving Nature? The Science and Ethics of Genetic Engineering*, Cambridge: Cambridge University Press, 1996.

Alan McHughen, *A Consumer's Guide to GM Food: from Green Genes to Red Herrings*, New York: Oxford University Press, 2000.

Matt Ridley, *Genome*, London: Fourth Estate, 1999.

John Harris, *Clones, Genes and Immortality: Ethics and the Genetic Revolution*, Oxford: Oxford University Press, 1998.

Philip Kitcher, *Lives to Come: the Genetic Revolution and Human Possibilities*, New York: Simon & Schuster, 1996.

Theresa Marteau and M. P. M. Richards (eds.), *The Troubled Helix – Psychosocial Implications of the New Human Genetics*, Cambridge: Cambridge University Press, 1996.

Dan W. Brock, Allen Buchanan, Norman Daniels and Daniel Wikler, *From Chance to Choice: Genethics and Justice*, Cambridge: Cambridge University Press, 2000.

Julian Morris (ed.), *Rethinking Risk and the Precautionary Principle*, Oxford: Butterworth Heinemann, 2000.

Onora O'Neill, *Autonomy and Trust in Bioethics*, Cambridge: Cambridge University Press, 2002.

Graeme Laurie, *Genetic Privacy: a Challenge to Medico-Legal Norms*, Cambridge: Cambridge University Press, 2002.

David B. Resnik, *The Ethics of Science: an Introduction*, London: Routledge, 1998.

Notes on the contributors

Dorothy Bishop is a psychologist who trained in Experimental Psychology at Oxford University, and in Clinical Psychology at London University, before taking a D.Phil. in Neuropsychology at Oxford. Her interest in genetics developed during the course of a research career investigating the nature and causes of language impairments in children, funded first by the Medical Research Council, and subsequently by the Wellcome Trust. She is currently a Wellcome Principal Research Fellow and Professor of Developmental Neuropsychology at Oxford University.

Malcolm Grant, a barrister, is Vice-chancellor of University College London since 2003. Previously, he was Pro-Vice Chancellor and Professor of Land Economy at Cambridge University, and a Fellow of Clare College. He is an environmental lawyer and in 2000 was appointed to be Chair of the Government's new 20-member Agriculture and Environment Biotechnology Commission, whose job is to provide strategic advice to the Government on the implications of biotechnology (including genetic modification) for agriculture and the environment. He was recently also appointed by the Government to lead the national public debate on GM, to assist in policy-making around issues such as the commercial growing of GM crops in the UK.

Sir Alec Jeffreys FRS was educated in Biochemistry and Genetics at the University of Oxford, and currently holds the positions of Professor of Genetics and Royal Society Wolfson Research Professor in the Department of Genetics at the University of Leicester. He is best known for his invention of genetic fingerprinting, though he has contributed extensively to other branches of human genetics, particularly in the fields of human DNA variation and mutation. He has a Knighthood for Services to Genetics, a Fellowship of the Royal Society, the Albert Einstein World of Science Award for 1996 and the Australia Prize. He has also received recognition outside the scientific community, being voted 'Midlander of the Year' for 1989 and being made an Honorary Freeman of the City of Leicester in 1993.

Sir Aaron Klug OM FRS was educated at the universities of Witwatersrand, Cape Town and Cambridge. He began as a medical student, transferred to science, and his Ph.D. at the Cavendish Laboratory was in Physics. He joined the MRC Laboratory of Molecular Biology in Cambridge in 1962, was the Director of the Laboratory from 1986 to 1996, and now continues as a retired worker, leading a research group on the structural biology of control of gene expression. In 1982, he was awarded the Nobel Prize in Chemistry for his development of crystallographic electron microscopy and his structural elucidation of biologically important nucleic acid – protein complexes. He was President of the Royal Society (1995–2000), and is a member of the Order of Merit. He was a colleague of Rosalind Franklin at Birkbeck College in the 1950s, soon after the time when her X-ray diffraction of DNA provided key information which allowed Watson and Crick to propose the double helical structure.

Torsten Krude was educated at the University of Konstanz. He joined the Wellcome/CRC Institute in Cambridge in 1994 as a post-doctoral fellow with support from the European Molecular Biology Organisation (EMBO) and the Royal Society. He is currently a University Lecturer in Cell Biology in the Department of Zoology at the University of Cambridge. His research focuses on the molecular mechanisms that regulate the replication of chromosomal DNA in human cells. He is a Fellow of Darwin College.

Ron Laskey FRS started his career in Oxford, followed by post-doctoral positions on the scientific staff of the Imperial Cancer Research Fund in London and the MRC Laboratory of Molecular Biology in Cambridge. There he discovered signals that direct proteins to the cell nucleus and invented a range of sensitive methods for detecting radio-isotopes. In 1983 he moved to become Charles Darwin Professor in the University of Cambridge, first in the Department of Zoology, then in the Wellcome/CRC Institute and now as Director of the MRC Cancer Cell Unit in the Hutchison/MRC Research Centre. His main interest has been how cells control DNA synthesis. He has developed cell-free systems that allow these processes to be studied in a test tube. Some of the proteins studied in this work are emerging as promising markers for the development of screening tests for the commonest cancers. He is a Fellow of Darwin College and a Fellow of the Royal Society. His work has been recognised by awards from several countries, including the Louis Jeantet Prize for Medicine. On a lighter note, he has written and recorded two albums called 'Songs for Cynical Scientists' and 'More Songs for Cynical Scientists'.

Baroness Onora O'Neill CBE FBA is Principal of Newnham College, Cambridge. She lectures in the Faculty of Philosophy and the Department of History and Philosophy of Science, and has written books and articles on ethics, bioethics, political philosophy and the philosophy of Immanuel Kant. She is a former member and chair of the Nuffield Council on Bioethics and the Human Genetics Advisory Commission, and chairs the Nuffield Foundation. She is a Member of the House of Lords, sits as a crossbencher and was a member of the Select Committee on Stem Cell Research.

Svante Pääbo obtained his education at the University of Uppsala. He was the founding Director of the Max Planck Institute for Evolutionary Anthropology in Leipzig, which opened in 1997, and is now Professor of Evolutionary Genetics, and Managing Director, there. He pioneered methods for the retrieval of DNA sequences from archaeological and palaeontological remains to elucidate the history of humans and Pleistocene mammals. By comparative DNA sequencing in humans, chimpanzees, gorillas and orangutans, he aims to gain a better understanding of the origin, time and early migrations of humans and their closest relatives.

Lord Robert Winston FCOG is Professor of Fertility Studies at Imperial College School of Medicine, London University, and Director of NHS Research and Development for Hammersmith Hospital. As a peer he takes the Government Whip (Lord Winston of Hammersmith since 1995) and speaks regularly in the House of Lords on education, science, medicine and the arts. He was recently Chairman of the House of Lords Select Committee on Science and Technology and is a board member of the Parliamentary Office of Science and

Technology. He regularly presents BBC science programmes including *Your Life in Their Hands*, *Making Babies*, *The Human Body*, *Secret Life of Twins* and, most recently, *The Superhuman* and *A Child of our Time*.

His contributions to clinical medicine include the development of gynaecological microsurgery in the 1970s and his team has established various improvements in reproductive medicine, subsequently adopted internationally, particularly in the fields of endocrinology, IVF and reproductive genetics.

His awards include the Cedric Carter Medal, Clinical Genetics Society, 1993, and the Victor Bonney Medal for contributions to surgery, Royal College of Surgeons, 1993. He was Gold Medallist for the Royal Society of Health in 1998, received the BMA Gold Award for Medicine in the Media in 1999, and, in the same year, the Faraday Gold Medal from The Royal Society.

Index

Numbers in italics refers to figures.